e-Business in Construction

e-Business in Construction

Edited by

Chimay J. Anumba and Kirti Ruikar

Foreword by

Professor Ronald McCaffer

WILEY-BLACKWELL

A John Wiley & Sons, Ltd., Publication

This edition first published 2008
© 2008 Blackwell Publishing Ltd

Blackwell Publishing was acquired by John Wiley & Sons in February 2007. Blackwell's publishing programme has been merged with Wiley's global Scientific, Technical, and Medical business to form Wiley-Blackwell.

Registered office
John Wiley & Sons Ltd, The Atrium, Southern Gate, Chichester, West Sussex, PO19 8SQ, United Kingdom

Editorial offices
9600 Garsington Road, Oxford, OX4 2DQ, United Kingdom
2121 State Avenue, Ames, Iowa 50014-8300, USA

For details of our global editorial offices, for customer services and for information about how to apply for permission to reuse the copyright material in this book please see our website at www.wiley.com/wiley-blackwell.

The right of the author to be identified as the author of this work has been asserted in accordance with the Copyright, Designs and Patents Act 1988.

All rights reserved. No part of this publication may be reproduced, stored in a retrieval system, or transmitted, in any form or by any means, electronic, mechanical, photocopying, recording or otherwise, except as permitted by the UK Copyright, Designs and Patents Act 1988, without the prior permission of the publisher.

Wiley also publishes its books in a variety of electronic formats. Some content that appears in print may not be available in electronic books.

Designations used by companies to distinguish their products are often claimed as trademarks. All brand names and product names used in this book are trade names, service marks, trademarks or registered trademarks of their respective owners. The publisher is not associated with any product or vendor mentioned in this book. This publication is designed to provide accurate and authoritative information in regard to the subject matter covered. It is sold on the understanding that the publisher is not engaged in rendering professional services. If professional advice or other expert assistance is required, the services of a competent professional should be sought.

Library of Congress Cataloging-in-Publication Data
e-Business in construction/edited by Chimay J. Anumba and Kirti Ruikar.
    p. cm.
  Includes bibliographical references and index.
  ISBN-13: 978-1-4051-8234-8 (hardback : alk.paper)
  ISBN-10: 1-4051-8234-2 (hardback : alk.paper) 1. Construction industry.
    2. Electronic commerce. I. Anumba, C.J. (Chimay J.) II. Ruikar, Kirti.
  HD9715.A2E28 2008
  381'.142—dc22
                                        2008006976

A catalogue record for this book is available from the British Library.

Set in 9.5/12.5 pt Palatino by Charon Tec Ltd (A Macmillan Company), Chennai, India
(www.charontec.com)
Printed in Singapore by Utopia Press Pte Ltd

1 2008

Contents

Contributors	xi
Foreword	xvii
Acknowledgements	xix
Abbreviations	xxi

1 Introduction — 1
Chimay J. Anumba and Kirti Ruikar
1.1 Context — 1
1.2 Structure of the book — 2
References — 4

2 Fundamentals of e-Business — 6
Kirti Ruikar and Chimay J. Anumba
2.1 Introduction — 6
2.2 e-Business and e-commerce definitions — 6
2.3 Taxonomy for e-business — 7
2.4 The four faces of e-business — 10
2.5 e-Business models — 11
2.6 e-Business in construction — 14
2.7 Summary and conclusions — 20
References — 21

3 e-Business: The Construction Context — 23
Kirti Ruikar, Chimay J. Anumba and Patricia Carrillo
3.1 Introduction — 23
3.2 e-Business and the construction business processes — 25
3.3 e-Business applications and end-user construction companies — 32
3.4 Summary — 40
References — 40

4 Organizational Readiness for e-Business — 42
Kirti Ruikar, Chimay J. Anumba and Patricia Carrillo
4.1 Introduction — 42
4.2 Methodology for e-readiness — 44
4.3 Review of readiness assessment models — 45
4.4 Verify end-user e-readiness using a diagnostic tool — 47
4.5 Verdict: System architecture and operation — 53
4.6 End-user case study — 56

	4.7 Conclusions and future work	60
	Acknowledgements	61
	Notes	61
	References	61
5	**Integrated Multi-Disciplinary e-Business Infrastructure Framework**	**65**
	Ihab A. Ismail and Vineet R. Kamat	
	5.1 Introduction	65
	5.2 Integrated construction e-business infrastructure framework	66
	5.3 The importance of e-construction infrastructure	67
	5.4 Summary and status of e-construction challenges	69
	5.5 Conclusions	77
	Note	77
	References	78
6	**The Role of Extranets in Construction e-Business**	**81**
	Paul Wilkinson	
	6.1 Introduction	81
	6.2 Defining construction collaboration technologies	81
	6.3 Uptake of construction collaboration technologies	83
	6.4 Benefits of construction collaboration technologies	86
	6.5 Human aspects of collaboration	91
	6.6 Moving beyond collaboration	99
	6.7 Conclusions	101
	References	102
7	**Agent-Based Systems: The Competitive Advantage for AEC-Specific e-Business**	**104**
	Esther A. Obonyo and Chimay J. Anumba	
	7.1 Introduction	104
	7.2 The current context	104
	7.3 Understanding agent-based systems	106
	7.4 A roadmap of agent-based systems in e-business	107
	7.5 APRON: An agent-based prototype system for AEC-specific e-business	109
	7.6 APRON's conceptual design	114
	7.7 The implemented APRON architecture	116
	7.8 Discussion and conclusions	117
	References	120
8	**The Role of e-Hubs in e-Commerce**	**123**
	Zhaomin Ren, Chimay J. Anumba and Tarek M. Hassan	
	8.1 Introduction	123
	8.2 e-Hub concept	124

8.3	e-Hubs' services	125
8.4	Engineering e-Hub	132
8.5	Engineering services	140
8.6	Problems and challenges	144
8.7	Conclusions	145
	Acknowledgements	147
	References	147

9 Web Services and aecXML-Based e-Business System for Construction Products Procurement — 149
Stephen C.W. Kong, Heng Li and Chimay J. Anumba

9.1	Introduction	149
9.2	The need for e-procurement of construction products	149
9.3	Existing e-business systems for construction products procurement	151
9.4	Limitations of existing e-business systems	153
9.5	The E-Union concept	154
9.6	Standardization of construction products information	155
9.7	The Web Services model of interoperable construction products catalogues	157
9.8	The E-Union Web Services prototypical implementation	160
9.9	Conclusions	164
	References	164

10 Using Next Generation Web Technologies in Construction e-Business — 167
Darshan Ruikar, Chimay J. Anumba and Alistair Duke

10.1	Introduction	167
10.2	The construction context	168
10.3	The need for the Semantic Web	169
10.4	The Semantic Web	172
10.5	Evolution of the Semantic Web in the construction sector	178
10.6	Semantic Web-based construction e-business	181
10.7	Summary	191
	References	192

11 Trust in e-Commerce — 195
Zhaomin Ren and Tarek M. Hassan

11.1	Introduction	195
11.2	Trust and trust building	196
11.3	Trust building in e-commerce	199
11.4	Conclusions	208
	References	209

12	**Legal Issues in Construction e-Business**	**211**
	Ihab A. Ismail and Vineet R. Kamat	
	12.1 Introduction	211
	12.2 Types of legal risks in construction e-business	212
	12.3 Contract formation, validity and errors	212
	12.4 Jurisdiction	213
	12.5 Privacy	214
	12.6 Authentication, attribution and non-repudiation	215
	12.7 Agency	216
	12.8 Conclusions	218
	References	219
13	**Knowledge Management for Improved Construction e-Business Performance**	**222**
	Charles O. Egbu	
	13.1 Introduction	222
	13.2 Knowledge management in context	223
	13.3 Exploiting opportunities in the fast-changing environment of e-business: A knowledge management perspective?	224
	13.4 Organizational challenges in using the internet to commercialize knowledge assets	226
	13.5 Knowledge assets employed by construction organizations in e-business initiatives	229
	13.6 Organizational readiness to launch a knowledge-business (k-business)	230
	13.7 Conclusions and recommendations	233
	References	233
14	**e-Commerce in Construction: Industrial Case Study**	**235**
	Tim C. Cole	
	14.1 Introduction	235
	14.2 Background	235
	14.3 A historic perspective	236
	14.4 e-Commerce implementation: Practical issues and benefits	239
	14.5 The first adopters	241
	14.6 Implementation issues: Case study examples	242
	14.7 Specific case study examples	245
	14.8 Summary	247
	References	247
15	**Assessment of e-Business Implementation in the US Construction Industry**	**248**
	Raymond R.A. Issa, Ian Flood and Bryce Treffinger	
	15.1 Introduction	248

	15.2	US construction industry	249
	15.3	e-Business assessment survey findings	254
	15.4	Conclusions	264
	References		264

16 Concluding Notes — **266**
Chimay J. Anumba and Kirti Ruikar

	16.1	Introduction	266
	16.2	Summary	266
	16.3	Benefits of e-business in construction	267
	16.4	Considerations in construction e-business implementation	268
	16.5	Future directions	270

Index — 273

Contributors

Chimay J. Anumba is Professor and Head, Department of Architectural Engineering, The Pennsylvania State University. He holds a PhD from the University of Leeds, and a higher doctorate (DSc) from Loughborough University. He is a Chartered/Professional Engineer with many years of industrial experience. Until recently, he was founding Director of the Centre for Innovative and Collaborative Engineering (CICE) and Professor of Construction Engineering and Informatics at Loughborough University. Professor Anumba's work has received support worth over £15 million and he has supervised over 31 doctoral candidates. He has over 400 publications and was recently awarded an Honorary Doctorate by Delft University of Technology. His Visiting Professorships exceed 10, including MIT and Stanford.

Patricia Carrillo holds a personal chair in Strategic Management in Construction in the Department of Civil and Building Engineering at Loughborough University. Professor Carrillo is also Programme Director for the department's MSc programmes in Construction Management and Construction Project Management. She has worked as a civil engineer with a range of clients, consultants and contractors. In 2002, she was awarded the prestigious Royal Academy of Engineering Global Award. This allowed her to have research secondments at the University of Calgary, Canada and University of Colorado, USA. Her area of expertise is in business performance and IT in construction.

Tim C. Cole is a Director at Causeway Technologies. He has been described as 'the construction industry's e-Envoy' by *Construction News* and *Construction Manager*. He has worked over 12 years to bring the benefits of e-business into mainstream construction. He has been involved in national and international e-business initiatives, including development of construction, banking and government data exchange standards. He is the author of *'Electronic Communication in Construction – Achieving Commercial Advantage'*, a book published in 2000, aimed at making e-business accessible to business professionals. He is also actively involved with the Network of Construction Collaboration Technology Providers (NCCTP) and is a founder member of the Hub Alliance.

Alistair Duke is a Principal Researcher within the Next Generation Web Research Group of British Telecommunications plc. He holds a

MEng from Aston University and a PhD in Collaboration Systems for Concurrent Engineering from Loughborough University. His primary interest is in Semantic Web and its application to the fields of Knowledge Management, Service Oriented Architecture and Business Systems. He was a member of the EU OnToKnowledge and Semantic Web-enabled Web Services (SWWS) projects and a work package leader on the EU DIP (focusing on applying Semantic Web Services to the Telecommunications industry) and SEKT project (where he led the Knowledge Access work package, developing end-user tools supported by Semantic technology).

Charles O. Egbu holds a personal Chair in Project Management and Strategic Management in Construction at the University of Salford, UK. Professor Egbu lectures at undergraduate and postgraduate levels in the areas of Construction Management, Project and Programme Management. With over 200 publications, his research interests are in innovation, knowledge management, e-business and construction project management.

Ian Flood received his PhD from the University of Manchester in 1986, developing parallel approaches to construction process simulation. He has held academic positions in Building and Civil Engineering at the National University of Singapore and the University of Maryland, and is currently at the Rinker School, University of Florida. His research has focused on the application of artificial intelligence methods of modelling to project planning and the simulation of the behaviour of engineered systems. Dr Flood has edited four books, has published over 100 refereed articles, and is highly cited, being the most frequently referenced author within the ASCE *Journal of Computing in Civil Engineering*.

Tarek M. Hassan is a Senior Lecturer and Director of the European Union Research Group in the Department of Civil and Building Engineering at Loughborough University. Dr Hassan's academic experience is complemented by 10 years of industrial experience with several international organizations. He has been leader and partner in 10 EU projects funded by the European Commission under the Information Society Technologies (IST) programme. He is also an appointed expert evaluator for research proposals submitted to the European Commission and appointed expert reviewer for ongoing EU research projects. He has over 95 publications and his areas of research include advanced ICTs, collaborative engineering, virtual enterprise business relationships, e-business, e-commerce and legal aspects of ICT.

Ihab A. Ismail is the Founder and Managing Director of Enovio, an international management consulting firm operating in several industries. After founding and developing Enovio, he focused his consulting practice on the capability development of firms with special interest in developing risk management capabilities within construction. He is currently a PhD student at the Construction Engineering and Management Department at the University of Michigan, USA. His research interests are in the areas of e-business and legal risk management in the construction industry. He holds a BSc in Civil Engineering from the American University in Cairo, and an MSc in Construction Management from the University of Michigan.

Raymond R.A. Issa is currently Rinker Professor and Director of Graduate and Distance Education programmes at the Rinker School, University of Florida. He received his PhD from Mississippi State University in 1982 and his JD in Law from the University of Memphis in 1985. He has held academic positions at various universities in the United States and his research has focused on the use of information technologies in construction and the legal aspects of construction management. Professor Issa has edited two books and five proceedings, has published over 120 refereed articles and he has chaired over 200 Masters and PhD Committees.

Vineet R. Kamat is an Assistant Professor in the Department of Civil and Environmental Engineering at University of Michigan. He holds a PhD and an MS in Civil Engineering from Virginia Tech, USA and a BE degree in Civil Engineering from Goa University, India. Dr Kamat's primary research interests are in advanced applications of computing for engineering and construction.

Stephen C.W. Kong is a researcher at Hong Kong Polytechnic University. His research interest is in the application of information technology in construction organizations. He has developed an e-business system for construction material trading that has been adopted in Hong Kong and China. Currently, he is working on adopting virtual prototyping technology in construction to improve construction planning through providing consultancy services to contractors. He also utilizes virtual prototyping technology to enhance student learning in construction technology.

Heng Li started his academic career in Tongji University in 1987. He then researched and lectured at the University of Sydney, James Cook University and Monash University before joining Hong Kong Polytechnic University. During this period, he worked with engineering design and construction firms and provided consultancy services to both private and government organizations in Australia, Hong Kong and China. He has led many funded research projects related to the innovative application and transfer of construction information technologies, and has published two books, in excess of 190 journal papers and presentations at numerous national and international conferences.

Esther A. Obonyo is an Assistant Professor at the University of Florida's Rinker School of Building Construction. She offers courses in construction engineering and productivity improvement. Her other research interests include continuous business improvement and the effective use of Information Technology to promote knowledge sharing. Dr Obonyo is a member of the Powell Centre for Construction and Environment and also sits in the College's Sustainability Committee. Previously, she worked as a business improvement analyst within the Balfour Beatty Group in the United Kingdom and a construction project manager in Nairobi, Kenya.

Zhaomin Ren is a Senior Lecturer at the School of Technology of Glamorgan University. Dr Ren's academic experience is complemented by 16 years of industrial experience as a design engineer, chief engineer, and project manager with several international organizations. He has been a key researcher on several national and international research projects in areas of construction management, project management and e-engineering including project procurement, contract management, project planning and control, collaborative engineering, risk management, value management, advanced ICT, virtual enterprise business relationships, and legal and trust aspects of ICT. He has over 32 publications and is also a senior consultant for China-Geo Engineering Corp. in Hong Kong and Sri Lanka.

Darshan Ruikar is a Senior Consultant at Ove Arup and Partners, an internationally renowned design and business consulting firm. He works within the Arup Major Projects group providing strategic advice on ICT implementation for Infrastructure and Building projects. He has also worked as a structural engineer both nationally and internationally. His industry experience is complemented by years of academic experience. He holds a PhD in Civil Engineering from University of Nottingham, an MSc in Structural Engineering from UMIST (Manchester) and a BE in Civil Engineering from Pune University in India. Dr Ruikar's area of expertise is in improving business performance through the use of advanced ICTs in the architectural, engineering and construction sector.

Kirti Ruikar is a Lecturer in Architectural Engineering in the Department of Civil and Building Engineering at Loughborough University. Dr Ruikar is an architect with experience of working on several building projects. Her industry experience is complemented by several years of academic experience. She holds a Bachelor of Architecture degree from University of Pune in India, following which she completed an MSc in Construction Innovation and Management and an EngD (Doctor of Engineering) from Loughborough University. Her primary research interests are in the strategic use of emerging technologies such as e-business, advanced ICTs and KM in construction. She is also Associate Editor of *Journal of Information Technology in Construction*.

Bryce Treffinger holds a Bachelor of Design degree from University of Florida, 2003 and a Master of Science in Building Construction from University of Florida, 2005. She is currently working for a large construction company in Jacksonville, Florida.

Paul Wilkinson is head of corporate communications at BIW Technologies Ltd, a leading provider of Software-as-a-Service applications to the architecture, engineering and construction sector. His construction experience dates back to 1987 and includes seven years with a firm of consulting engineers, four years with a major contractor, and two years running his own consultancy business. He is the author of *'Construction Collaboration Technologies: The Extranet Evolution'*, published in 2005. He is also actively involved with the Network of Construction Collaboration Technology Providers (NCCTP) and Constructing Excellence.

Foreword

The construction industry is frequently described as fragmented by its critics; however, disseminated would be a better description. For each construction project a whole new organization is created involving the client, designers, contractors, sub-contractors, material suppliers, plant hire companies, government, local authorities and agencies such as the environmental agency, Health and Safety Executive and many others. Each 'new' and 'transient' project organization is, in fact, a virtual organization or enterprise. The communications between the parties both contractual and non-contractual are of a scale nearing unmanageable proportions. Many of the inefficiencies identified in the reports of Latham and Egan were rooted in the communications processes.

The developments in information and communication technologies are changing the way that many industry sectors conduct their business operations and offer the solution to the construction industry's communications in the project delivery process. Not only is there now a growing dependence on electronic communications by construction sector, some companies and researchers in the sector are actually leading the way in the development of e-business for the benefit of our industry as well as other sectors. However, many companies and organizations still have difficulties in maximizing the opportunities offered by conducting business electronically.

e-Business has much to offer the construction sector, as it directly addresses the issues that a disseminated industry has to deal with – distributed collaboration, electronic sourcing and purchasing of products and services that meet well-defined requirements, globalization, need for improved efficiency and timely delivery (amongst others).

This book is a clear guide to e-business in the construction industry and will raise the awareness and knowledge of the industry. The authors cover a wide range of issues – basic definitions and fundamental concepts, the construction industry context for e-business, the role of emerging information and communication technologies, socio-technical and legal aspects of e-business implementation, and industry perspectives from both Europe and North America. Practical examples from construction and other industry sectors are used throughout to illustrate the various aspects of e-business. The challenges in implementing e-business are clearly articulated and the ensuing benefits highlighted.

It is widely accepted that e-business is the way to conduct construction business in the 21st century. It is the means available to companies to continuously improve efficiency and effectiveness in serving their clients' needs and in delivering a return to their shareholders. The development of e-business is upon us, now affecting the short and medium term but it is also the way business will develop over the long term.

This book provides practical guidance on how this can be done and I strongly recommend it to all participants in the construction process. It is based on sound research, scholarship and practical experience, and I consider it is essential reading for both construction researchers and industry practitioners.

Professor Ronald McCaffer, FREng
December 2007

Acknowledgements

We would like to thank the various colleagues, contemporaries and friends who have contributed towards our understanding and knowledge of the role of e-business in the construction sector. This book is a combined effort of several authors who have spent a considerable amount of time and effort in sharing their experiences and knowledge. We are immensely grateful for their contributions. Jo Brewin and Sara Cowin provided administrative assistance. We would also like to acknowledge the contribution of the various agencies and organizations that funded the research projects and initiatives on which this book is based. We owe special thanks to our families, for their continued love, support and encouragement that made this book both possible and worthwhile. Last but not the least, we must thank our young ones for the much needed distractions from work.

Professor Chimay J. Anumba
Dr Kirti Ruikar

Abbreviations

A2A	Administration-to-Administration
A2B	Administration-to-Business
A2C	Administration-to-Consumer
AEC	Architecture, Engineering and Construction
B2A	Business-to-Administration
B2B	Business-to-Business
B2C	Business-to-Consumer
BCP	Basic Collaboration Platform
C2A	Consumer-to-Administration
C2B	Consumer-to-Business
C2C	Consumer-to-Consumer
CAD	Computer Aided Design
CITE	Construction Industry Trading Electronically
CRM	Customer Relationship Management
DRM	Digital Rights Management
DSL	Digital Subscriber Lines
EAI	Enterprise Applications Integration
e-HUBs project	e-Engineering Enabled by Holonomic and Universal Broker Services
ENR	Engineering News Record
ESPs	Engineering Service Providers
EU	European Union
GUI	Graphical User Interface
ISP	Internet Service Provider
IT	Information Technology
OCS	Online Contracting System
OECD	Organisation for Economic Cooperation and Development
PKI	Public Key Infrastructure
PMIs	Privilege Management Infrastructures
PP	Project Planning
PPM	Project Planning Model
ROI	Return on Investment
SCM	Supply Chain Management
SMEs	Small and Medium-Sized Enterprises
SOAP	Simple Object Access Protocol
UDDI	Universal Description, Discovery, Integration
UETA	Uniform Electronic Transaction Act

UN/EDIFACT	Electronic Data Interchange for Administration Commerce and Transport
UPS	Uninterrupted Power Supply
VAT	Value-Added Tax
VHS	Video Home System
WF	Workflow
WfMS	Workflow Management System
WSDL	Web Service Description Language
WSIL	Web Service Inspection Language
XML	eXtensible Markup Language

1 Introduction

Chimay J. Anumba and Kirti Ruikar

1.1 Context

The rapid expansion of the Internet has transformed the way in which people and businesses communicate and interact. It has revolutionized the way in which information is stored, exchanged, viewed, and manipulated. This has opened up new opportunities for businesses, which were almost inconceivable before, as it is now possible to conduct business transactions on a global basis, within a relatively short span of time, and at a fraction of the amount it would have cost previously (using conventional methods).

Businesses have recognized the possibilities the Internet has to offer and hence the need to adopt new business processes. This has had some immediate consequences but there is a need for businesses to assess and rethink their existing processes and working methods in order to make the most of the available opportunities. In some cases, this will only require small, incremental changes; while in others, a radical transformation from staid, traditional methods is required to benefit from the new technologies. Such changes can prompt businesses to improve their traditional business processes, innovate their products and services, and develop strategies that are flexible enough to incorporate new technologies as they emerge. The increase in electronic ways of conducting business has had a major impact on virtually every business sector. The construction industry is no exception.

The construction industry is an ideal sector for the adoption of e-business, given the number of disparate organizations involved in the project delivery process. Each construction project has a supply chain that includes architects, engineers, project managers, contractors, quantity surveyors, sub-contractors, materials suppliers, and self-employed professionals or artisans. Increasingly, these supply chain members are geographically distributed and rely on electronic communications for business transactions and collaboration. This makes the use of e-business tools inevitable.

In the light of the above, the use of e-business tools in construction has been on the increase (Berning and Diveley-Coyne, 2000). The benefits of using these tools on construction projects have been documented in several publications (Anumba and Ruikar, 2002; Berning and Flanagan, 2003). However, the use of these tools is still not ubiquitous within the industry. For those construction companies that seek to adopt e-business tools there is a need to undertake an analysis of their business processes and working methods to ensure a productive and beneficial implementation of these tools.

This book is concerned with the implementation of e-business in the construction industry. In this context, e-business is broadly defined as the conduct of construction business by electronic means. The term, e-business, as used in this book, should therefore not be considered synonymous with narrow definitions of e-commerce as the process of buying and selling goods and services online (Laudon and Laudon, 2002; Unisys, 2004). In this regard, the issues and applications in this book cover both 'simple' buying and selling transactions as well as the wider aspects of doing business by electronic means. This makes the book essential reading for all stakeholders in a construction enterprise.

1.2 Structure of the book

This book consists of chapters that describe current developments and future directions in the theory and application of e-business in the architecture, engineering and construction (AEC) sector. The book reflects an effort from an international network of construction researchers investigating different aspects of e-business in construction.

The chapters, each of which is briefly described below, can be broadly grouped into four areas:

(1) Introduction to e-business and the construction context (Chapters 2–5).
(2) Technological support for e-business in construction (Chapters 6–11).
(3) Socio-technical issues in e-business (Chapters 11–13).
(4) Industrial case studies (Chapters 14–15).

Chapter 1 introduces the book, its context and contents. It explains the use of the term 'e-Business' and discusses the structure of the book, particularly the relationships between the various chapters.

Chapter 2 focuses on the fundamentals of e-business. It presents a general taxonomy for e-business and discusses the various classifications in details. It also discusses trends and developments in e-business both within and outside the construction industry. e-Business models and the general benefits of e-business are also covered, drawing on construction industry examples, where appropriate.

Chapter 3 discusses the construction context for e-business and draws on case studies involving both e-business product suppliers and construction end-user companies. The potential for business process re-engineering based on the implementation of e-business is highlighted and some of the associated benefits outlined.

Organizational readiness for e-business forms the focus of Chapter 4. It is argued that for construction sector organizations to reap the benefits of e-business they must be 'ready' with respect to four critical aspects of their operations – technology, management, process, and people. VERDICT – a tool specifically developed to help organizations measure their e-readiness is also described in detail.

Chapter 5 is concerned with an e-business infrastructure for multi-disciplinary teams. It argues that such an infrastructure extends beyond the physical hardware and routers that supports the network connections enabling e-business transactions. It proposes a framework for an integrated infrastructure that covers business process, trading models, network hardware and software infrastructure, software interoperability standards, legal standards, and related issues.

Chapter 6 discusses the role of extranets in construction e-business. It describes how these systems are expanding beyond the core area of document/drawing management towards support for various key AEC business processes. It also explores the current slow adoption of technology adoption, the reasons for this, and the human issues to be addressed.

Chapter 7 describes how agent technology can make AEC-specific e-business more effective and more efficient. The chapter provides an overview of agent technology and the vision for agent-based e-business, and presents the conceptual design of a prototype system that demonstrates the suggested approach. It also emphasizes the need to have realistic expectations when deploying agent-based systems in commercial settings.

Chapter 8 presents the basic concept of an electronic hub (e-Hub), and its key characteristics and services. The focus is on an engineering e-Hub – a novel concept developed by the EU funded e-Hubs project. The chapter describes the e-Hub developed as being able to offer a systematic approach to collaborative project planning through basic collaborative engineering services and collaborative workflows.

Chapter 9 introduces the concept of electronic union (E-Union), which integrates the information and services provided by different e-business systems for construction products procurement. It describes the underlying technologies (such as Web Services and XML) for implementing E-Union. The last part of the chapter describes the Web Services and aecXML-based framework of E-Union and a prototypical implementation.

Chapter 10 explores the potential of emerging Web technologies (such as the Semantic Web) and describes the potential for the future application of Semantic Web technologies in construction e-business. Using examples, it argues that the use of Semantic Web technologies will, in future, offer considerable benefits in terms of enhanced e-business, project management, knowledge management, supply chain management, integration of distributed applications and services, and improved project delivery processes.

Chapter 11 discusses the issue of Trust, which is seen as a cornerstone of e-business. It reviews fundamental concepts associated with Trust and specifically focuses on trust building in e-business environments. It also discusses a number of trust building models and outlines some of the technologies that are intended to foster trust building in e-commerce by improving online security and privacy.

The legal framework for construction e-business is the subject of Chapter 12. The legal issues and risks associated with various aspects of e-business are described in detail, and guidance provided on how to avoid major problems. The chapter classifies the risks under contract formation, validity and errors, jurisdiction, privacy, authentication, attribution, non-repudiation, and agency.

Chapter 13 documents the key contributions which knowledge management (KM) can make to effective e-business initiatives and the factors that promote and inhibit the contributions of KM to e-business. It outlines the main strategies and processes needed to fully exploit the benefits of KM, and argues that to be successful in the e-business environment, construction sector companies need to become knowledge-based businesses.

Chapter 14 addresses the practical implementation of e-business systems within the UK construction industry. Drawing on case studies, it discusses the key issues that often arise in implementing e-business. It provides tips on how these problems can be avoided or overcome, and argues that technological solutions must be flexible enough to support the business objectives and facilitate innovation.

In Chapter 15, a survey of e-business implementation in the US construction industry is presented. It explores the extent to which the US construction industry has redefined its way of doing business based on e-business, and discusses the future implementation potential for various industry participants: general contractors, design firms, suppliers, and sub-contractors in general. It focuses on the integration of e-business into construction project management systems by general contractors.

Chapter 16 summarizes the main issues raised in various chapters of the book. It highlights the benefits, barriers, and practical implementation issues in the adoption of e-business in the construction industry.

The chapters in the book can be read in any sequence. However, it is important that readers appreciate the need for a balance between the technological and human/organizational aspects of e-business implementation in the construction industry.

References

Anumba, C.J. and Ruikar, K. (2002) E-commerce in construction: trends and prospects. *Automation in Construction,* **11**, 265–275.

Berning, P.W. and Diveley-Coyne, S. (2000) *E-commerce and the Construction Industry: The Revolution is Here*. Thelen Reid and Priest LLP. Available at http://www.constructionweblinks.com (accessed November 2007).

Berning, P.W. and Flanagan, P. (2003) *E-commerce and the Construction Industry: User Viewpoints, New Concerns, Legal Updates on Project Websites, Online Bidding and Web-based Purchasing*. Thelen Reid and Priest LLP. Available at: http://www.constructionweblinks.com (accessed November 2007).

Laudon, K.C. and Laudon, J.P. (2002) Management information systems. In *Managing the Digital Firm*, 7th edition. Prentice-Hall Inc., NJ, USA.

Unisys (2004) *2000 Annual Report: Glossary*. Available at: http://www.unisys.com/annual/annual2000/glossary/ (accessed November 2007).

2 Fundamentals of e-Business

Kirti Ruikar and Chimay J. Anumba

2.1 Introduction

This chapter discusses the fundamentals of e-business including the definitions of e-business and e-commerce. It presents taxonomy for e-business and the four faces of e-business. It discusses some e-business models and reviews e-business trends in construction including the barriers and enablers for e-business in construction.

2.2 e-Business and e-commerce definitions

There are a range of definitions of e-business. Damanpour (2001), for example, defines e-business as any 'net' business activity that transforms internal and external relationships to create value and exploit market opportunities driven by new rules of the connected economy. The Gartner Advisory Group (Damanpour, 2001), a research and advisory services firm, describes e-business in terms of a quantity rather than an absolute state of a company. They consider a business an e-business to the degree that it targets the market opportunities of conducting business under new electronic channels, which revolve around the Internet. This is an acknowledgement that e-business comes in many forms and can be implemented to a very small or a large degree. It is also an acknowledgement that the 'Internet' is an essential component of an e-business strategy. Laudon and Laudon's (2002) definition of e-business, as the use of the Internet and other digital technology for organizational communication, coordination and the management of the firm, encompasses these different adaptations. In the broadest possible terms, however, e-business is an electronic way of doing business. The fact that the value proposition of e-business includes the creation of new market opportunities through electronic channels, should not be ignored as these electronically channelled market opportunities enable companies to lower transaction costs, reduce delivery times, improve customer services, and add convenience (Damanpour, 2001).

This book is concerned with the implementation of e-business in the construction industry. In this context, e-business is defined broadly as the conduct of construction business by electronic means. This fits with broad definitions of the term 'e-commerce' exemplified by the definitions below:

- The Organization for Economic Cooperation and Development (OECD): 'The electronic exchange of information that support and govern commercial activities including organizational management, commercial management, commercial negotiations and contracts, legal and regulatory frameworks, financial settlement arrangements and taxation' (OECD, 1999).
- Learnthat (2004): e-Commerce is not just about buying and selling online, but also includes all forms of business activities that are conducted over the Internet (e.g. the business-to-business flow of information between companies or within a company, communication between businesses, online advertising, etc.).
- Kalakota and Whinston (1997): e-Commerce at its grass root level can be described as an electronic method of doing business, typically over the Internet. Broadly defined, however, 'e-commerce is a modern business methodology that addresses the needs of organizations, merchants and consumers to cut costs while improving the quality of goods and services, and increasing the speed of service delivery'.

Thus, the term, e-business, as used in this book should not be considered synonymous with narrow definitions of e-commerce as the process of buying and selling goods and services online (Laudon and Laudon, 2002; Unisys, 2004).

2.3 Taxonomy for e-business

e-Business can be broadly divided into the following categories as illustrated in Figure 2.1:

- Business-to-Business (B2B)
- Consumer-to-Consumer (C2C)
- Administration-to-Administration (A2A)
- Business-to-Consumer (B2C) or Consumer-to-Business (C2B)
- Business-to-Administration (B2A) or Administration-to-Business (A2B)
- Consumer-to-Administration (C2A) or Administration-to-Consumer (A2C)

2.3.1 *Business-to-Business (B2B)*

Business-to-Business (or B2B as it is commonly referred to) is an electronic means of carrying out business transactions between two or more businesses. B2B incorporates everything from manufacturing to service providers.

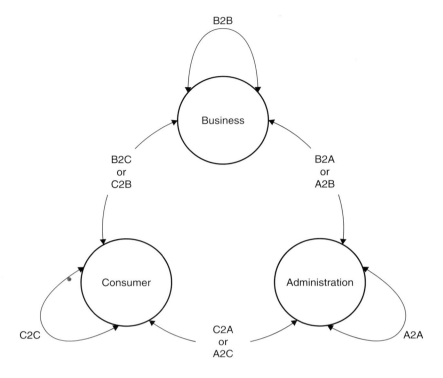

Figure 2.1   e-Business taxonomy

There are several examples of B2B models. Using B2B a company can leverage the Internet to place orders electronically, receive electronic invoices and make electronic payments. Other examples include project extranets, e-Hubs and e-commerce applications which are covered in Chapters 6, 8 and 14 of this book, respectively.

2.3.2   *Consumer-to-Consumer (C2C)*

Examples of C2C business models include, consumer e-auctions and blogs. In a C2C business model, although there may be no financial transaction, there is still an exchange of value and these are economic activities and could be referred to as peer-to-peer (Timmers, 2000). Blogs, for example, have led to the development of news C2B and C2C applications by presenting the opportunity and tools for virtually anyone to express their views easily and to communicate these globally and inexpensively. For instance, Nano-publishing is an application of C2C (and C2B) schemes using low-cost online publishing techniques such as blogging (writing weblogs) to target a specific audience. Additionally, Podcasting, video casting, and other blog-related technologies help to provide opportunities to develop new economic systems and to generate alternative revenues.

2.3.3 Administration-to-Administration (A2A)

Using the A2A model, government departments can nationally and or internationally communicate and exchange classified information through dedicated portals. Typical examples include the national DNA database and other policing information.

2.3.4 Business-to-Consumer (B2C) or Consumer-to-Business (C2B)

In a B2C model, commercial transactions are between an organization and the consumers (Chaffey, 2002). When applied to the retail industry, for instance, a B2C process will be similar to the traditional method of retailing, the main difference being the medium used to carry out business – the Internet. Such a method of carrying out business transactions assumes that the consumer has access to the Internet. By selling direct to customers or reducing the number of intermediaries, companies can achieve higher profits while charging lower prices (Laudon and Laudon, 2002). This removal of intermediary organizations or business process layers is termed disintermediation. Some examples of the B2C category include Amazon.com and eBay.

C2B on the other hand, is a business model in which consumers offer products and services to companies at a cost. This business model is a reversal of traditional business model where companies offer goods and services to consumers. Online surveys such as Surveys.com, and SurveyMonkey, are typical examples of C2B models, where individuals offer the service to reply to a company's survey and in return the company pays the individual for their service.

2.3.5 Business-to-Administration (B2A) or Administration-to-Business (A2B)

The B2A category covers all transactions that are carried out between businesses and government bodies using the Internet as a medium. This category has steadily evolved over the last few years. An example of a B2A model, is that of Accela.com, a software company that provides round the clock public access to government services for asset management, emergency response, permitting, planning, licensing, public health, and public works.

A2B is an electronic means of providing business-specific information such as policies, regulations directly to the business. A typical example of the A2B category is construction e-tendering solutions that enable potential construction stakeholders to bid for government-led projects such as the 2012 London Olympics, using online tendering tools.

2.3.6 Consumer-to-Administration (C2A) or Administration-to-Consumer (A2C)

The C2A and A2C categories have emerged in the last decade. C2A examples include applications such as e-democracy, e-voting, information about public services and e-health. Using such services consumers can post concerns, request feedback, or information (on planning application progress) directly from their local governments/authorities.

A2C provides a direct communication link between governments (e.g. local authority) and consumers. The HM Revenue and Customs Website, for example, allows consumers to directly file tax returns using a secure Website. Other examples are those of local council and civic service Websites that inform the general public about community events, road closures, and other activities that impact the community and public services.

2.4 The four faces of e-business

Damanpour (2001) discusses the four faces of e-business, which were originally identified by the Gartner Advisory Group each of which looks at e-business from a different perspective (Figure 2.2). The four faces include the following:

(1) *Face 1: Business and Financial Models Perspective* – This face focuses on the business model and opportunities that operate as an electronic entity. Financial considerations such as reduced costs and operations efficiency are primary considerations. Such a model regards technology as an enabler of the business opportunity and sometimes requires changes in the corporate culture, financial accounting guidelines, and the corporate image. The model can be used for an existing company (brick-and-mortar), a new spin-off from brick-and-mortar, or a small unknown start-up (e.g. amazon.com when it was first launched).

(2) *Face 2: Relationships* – This face looks at e-business from a relationship perspective as new relationships and collaborations are created and forged in e-business to enter new markets or enhance customer, supplier and business relations. Some examples of the relationship perspective are customer relationship management (CRM), supply chain management (SCM) and Infrastructure management. For example, the traditional ordering and invoicing processes can be managed electronically. Electronic marketplaces, catalogues and bidding systems, and Internet searches can transform business demeanour, accelerate business activities, increase global competition, create global logistics networks, provide improved customer relationships, cost-effective services, and speed up goods and information delivery down the entire supply chain.

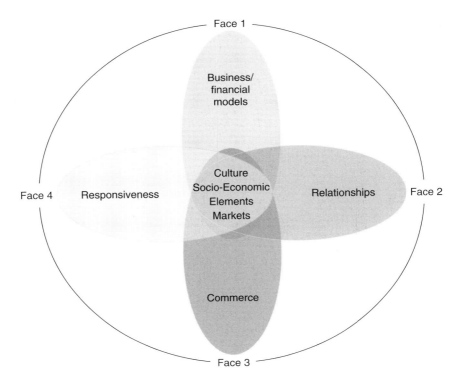

Figure 2.2 **Four faces of e-business.** (*Source*: Adapted from Damanpour, 2001)

(3) *Face 3: Commerce* – This face focuses on electronic buying and selling, which requires the development of systems, services, models, and relationships to support effective buying and selling. Face 3 overlaps other three faces and emphasizes the importance of technology to business success and customer demands by leveraging the capability of the Internet to reach global buyers around the clock.

(4) *Face 4: Responsiveness* – This face is centred on the efficiency and timing of business transactions. Responsiveness, in e-business terms, means reducing the time between a business request and its fulfilment by increasing efficiency of the delivery of processes and their supporting computing systems for seamless operations to provide fulfilment. For example, the direct connection of a rent-a-car automobile request system to insurance companies results in improved efficiency, reduction of errors, and hence customer satisfaction.

2.5 e-Business models

The Internet has changed the ways in which companies manage businesses. Traditionally, for example, if a car buyer required information about a car, it would mean several visits to car dealer showrooms (Figure 2.3). Cost comparison, in such instances was complicated as the potential buyer would invest considerable time and effort to visit each dealer. Such models relied on the physical location of the supplier and buyer and quite often businesses were 'localized' in their clientele. The e-business models extend the traditional boundaries beyond the physical location to potentially global markets (Figure 2.4). Onestopcaradvice.co.uk, for example, allows users to

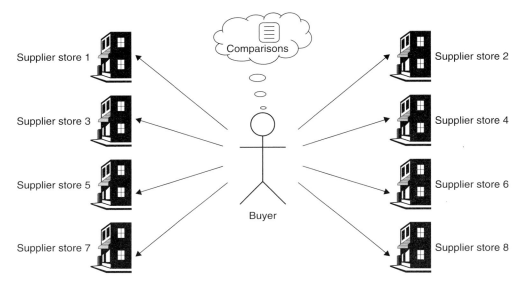

Figure 2.3   Traditional buyer–supplier environment

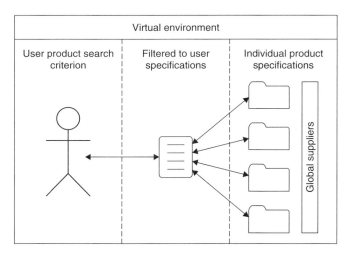

Figure 2.4   Buyer–seller relationship in an e-business environment

search for used and new cars from a wide selection of franchised car dealers and second hand car dealers across the United Kingdom. So the buyer can now compare and contrast car prices and specifications and shortlist/select those that best fit their specifications.

Laudon and Laudon (2002) present a useful classification of business models for e-business. This is summarized in Table 2.1 and includes examples in each category. These business models are more appropriate for some of the categories of e-business than others as is evident from table below. Organizations have to determine, which model is best suited to their business needs.

Table 2.1 Examples of e-business models

e-Business model type	Description
Virtual Storefront	Sells physical goods or services online instead of through a physical storefront or retail outlet. Delivery of non-digital goods and services takes place through traditional means (Amazon.com, Wine.com).
Market place concentrator	Concentrates information about products and services from multiple providers at one central point. Purchasers can search, comparison-shop, and sometimes complete the sales transaction (DealerNet.com, dooyoo.co.uk).
Online exchange	Bid–ask system where multiple buyers can purchase from multiple sellers (E-Steel Stockhouse.com).
Information brokers	Provide product, pricing, and availability information. Some facilitate transactions, but their main value is the information they provide (pricerunners.co.uk).
Transaction brokers	Buyers can view rates and terms, but the primary business activity is to complete the transaction (Ameritrade).
Auction	Provides electronic clearinghouse for products where price and availability are constantly changing, sometimes in response to customer actions (eBay).
Reverse auction	Consumers submit a bid on multiple sellers to buy goods or services at a buyer-specified price (Hedgehog.com, R3versebid.com).
Aggregator	Groups of people who want to purchase a particular product sign up and then seek a volume discount from the vendor.
Digital product delivery	Sells and delivers software, multimedia, and other digital products over the Internet (Hallmark.com).
Content provider	Creates revenue by providing content. The customer may pay to access the content, or revenue may be generated by selling advertising space or by having advertisers pay for placement in an organized listing in a searchable database.
Online service provider	Provides service and support for hardware and software users (xdrive.com).
Application service provider	Operates software at its data centre which customers access online under a service contract (BIW Technologies, 4Projects).

(*Continued*)

Table 2.1 (*continued*)

e-Business model type	Description
Virtual community	Provides online meeting place where people with similar interests can communicate and find useful information (Wikipedia, Geocities).
Portal	Provide initial point of entry to the Web along with specialized content and other services (Yahoo, Lycos).
Syndicator	Aggregates contents or applications from multiple sources and resells them to other companies (Thinq, Screaming Media).

Source: Adapted from Laudon and Laudon, (2002).

2.6 e-Business in construction

The uptake of e-business in the construction industry has been relatively limited and ineffective as compared to other engineering sectors such as the automotive or the aerospace industry. One of the reasons for this is the fragmented nature of the construction industry and the one-off nature of the product. A construction project is a complex activity involving several participants; for example, the client, architect, structural engineer, fabricator and the contractor. It is a team effort, involving several, inter-organizational activities and dialogue. Traditional communication and document exchange models were often manual and hence slow. The traditional means of communication involves producing numerous paper copies of documents and drawings. Management of these loose documents is often very time-consuming and tedious. Libraries of documents need to be maintained to effectively access data as and when required by the user. 'A lack of a clear audit trail causes delays in communicating with other members of the team' (Needleman, 2000). The reliance on third parties such as courier services, can sometimes lead to delays. There is also a high amount of added expense incurred in the delivery of project documents to project members who are geographically distributed. e-Business has the potential to overcome some of the process and communication inefficiencies. Some of the common construction e-business trends include:

- *Service promotions*: The Internet is being used to promote companies by the dissemination of company service information. Architects, designers, fabricators contractors, and other members of the construction sector are using the Web to promote their companies and inform potential clients of their services. The main idea behind such a site is to promote the company and its services to its targeted market.
- *Product promotion*: The Internet is used for the purpose of increasing product sales through online promotion. Product promotion is done either through an independent Website or through an online vendor. Such a product promotion site displays all product and material specifications

that can include manufacturer and supplier details, product availability, quality assurance, cost and mode of delivery. This information is stored categorically and is regularly updated.

- *e-Procurement through Web directories and search engines*: Some of the principal methods of locating information on the Web are with the help of search engines, Web directories, and broadcast or 'push' technology (Laudon and Laudon, 2002). Search engines search documents for specified keywords and return a list of the documents where the keywords are found. The Web can be used as a tool to procure information about construction-related suppliers and their products. Several Websites provide a search tool for the user to access varied information about the construction industry. Information may be varied and can range from items such as jobs and products to specific information about bidding processes.
- *Project management*: Some Websites are designed to streamline the construction business process. These sites look into how the Internet can be used to improve and integrate the process of design and management of a construction project. Such a project management site may yield several benefits to its users. It can result in speeding up the process of communication between different parties involved in a construction project and thus avoid any unnecessary delays that are often a direct result of miscommunication.
- *Project collaboration*: The Web can be used as a tool to facilitate online collaboration for project partners, which allows project partners to collaborate and communicate with each other in real time. The concept of online collaboration defies the boundaries of time and geography and allows construction stakeholders to among other things, exchange ideas, and make comments no matter where they are located. There are several online project collaboration tools available for construction project teams.
- *Online tendering*: The Internet has now made it possible to have online tendering services. With this facility it is possible to provide tendering information online along with project specifications.

2.6.1 *e-Business enablers*

The emergence of e-business has revolutionized the way in which companies conduct business. Some of the advantages of previously discussed e-business applications in the construction sector can be summarized as follows:

- *Service/product promotions*: Using the Internet to promote a company or its products can facilitate a reduction in advertising and marketing costs, provision of company information (products and services) through a Web presence, easy access to target audiences from the construction sector, and transparency with customers.

- *e-Procurement through search engines and Web directories*: Internet-based construction search engines can allow several advantages. These include quicker access to construction-related information, up-to-date product and industry information, simplified procurement business processes, cost savings through disintermediation, and quicker product comparison in terms of price and quality.
- *Project management/online project collaboration*: Online collaboration tools can facilitate easier management of construction projects, easier access to project information from anywhere at anytime, faster transaction time, better transparency in the exchange of project information, better collaboration between construction project partners, time savings for communication of project information, savings on project cost, and streamlined construction business processes.
- *Other benefits*: In addition to the benefits cited above using e-business applications in construction can reduce paperwork, reduce re-keying of information (thereby reducing errors), and provide wider market reach.

Zou and Seo (2006) discuss how the use of e-business technologies has generated primary and secondary effects in the construction processes, where:

- Primary effects are more efficient information-related activities such as creation, retrieval, delivery of information and effective communication, which assisted increasing productivity.
- Secondary effects are activities related to efficient material-handling in information processing through the use of information technology where it contributes to inventory reduction and decrease in the number of rebuilding with accurate design information and production of energy efficient buildings.

By using e-business in construction, stakeholder organizations can benefit from lower transaction costs, reduced staffing requirements, shorter procurement cycles, decreased inventory levels, higher level of transparency, provision of information on demand which promotes more frequent and intense use of it, connection to operations across organizational boundaries, enlargement of the span of effective control and coordination, improvement in the quality of decision-making processes; and enhanced communication and collaboration between supply chain members/organizations. Some of the potential benefits of using e-business in construction organizations are highlighted in Table 2.2.

2.6.2 Barriers to e-business

As is true with most technologies, there are challenges associated with e-business adoption, which need to be recognized, identified, and addressed. Doing so will improve public confidence in adopting

Table 2.2 Potential benefits of e-business to construction organizations

Construction discipline	Benefits of e-business
Clients/owners/developers	• Improved project efficiency • Reduced construction costs • Reduced chance of errors and need for rework • Compressed construction programme • Improved transparency
Designers	• Reduced re-work due to improved information and document sharing (e.g. CAD files) • Time savings • Improved communication and document management • Increased accuracy and speed of specification
Contractors and sub-contractors	• Lower administration and communication costs • Tendering and procurement efficiencies • Time savings • More project control and security • Enhanced project communication
Builder merchants	• Lower inventory and real estate costs • Lower cost of serving customers
Manufacturers	• Reduced channel costs • Improved access to information • Cost-effective access to actively purchasing and specifying customers
Material suppliers	• Advanced stock checking for planning and preparing material delivery. • Better management of inventory • Faster and effective dissemination of up-to-date product/service information to clients • Reduced advertising and promotion costs

Source: Adapted from Zou and Seo, (2006).

e-business. The barriers to e-business can be classified into two categories, namely generic barriers that are common to all industry sectors, including the construction sector and construction-specific barriers.

Generic barriers to e-business

The general barriers to e-business mainly fall into three categories, namely issues related to infrastructure, trust and reliability, and regulatory issues. Within each of these categories there are issues that need to be considered for governments, businesses and consumers alike (Thorbjornsen and Descamps, 1997).

- *Infrastructure*: The Internet is a global phenomenon, however, the telecommunications infrastructure of several developing countries is not

sufficiently developed to handle the advances in e-business technologies and to compete at par with their developed counterparts.
- *Trust and reliability*: Confidentiality of data must be maintained, as data must not be visible to 'eavesdroppers'. It is also important that communicating parties are able to authenticate the identity of the other party, and know when data integrity has been compromised. In addition, once an exchange is complete, it must be possible to prove that a transaction has taken place; and
- *Regulatory issues*: Companies see unclear regulatory issues (such as tax issues, legal issues and ethical issues) as major deterrents in adopting e-measures. e-Business analysts suggest that security is the most important factor that can inhibit companies from adopting e-business. There are several governments' initiatives that are aimed at combating these concerns. The US Government, for example, addressed these issues concerning global e-business in a paper by Clinton and Gore (1997). Some of the issues addressed in this paper include:
  ○ Financial issues such as customs, taxation, and electronic payments.
  ○ Legal issues such as 'Uniform Commercial Code' for e-business, intellectual property protection, and privacy/security issues.
  ○ Market access issues (e.g. telecommunications infrastructure and information technology).

Construction-specific e-business barriers

There are several factors that have limited the uptake of e-business in construction including the high cost of initial investment associated with building the necessary infrastructure and training of personnel, quantifying the return on investment (ROI), security of data in online transactions, integration with legacy systems and interoperability of distributed software applications over the Internet (Ugwu *et al.*, 2000). Furthermore the construction industry operates using 'arms length contractual relationships' that does not encourage unnecessary risk (Lewis, 1999). For most construction projects, teams are formed for the duration of the project and these last only as long as the project itself. This temporary nature of relationships provides little incentive for investing into innovative technologies such as e-business. Another major barrier to the implementation of e-business in the construction industry relates to the investment justification for construction firms especially SMEs. Elliman and Orange (2000) state that SMEs simply do not have the capital needed to implement e-business technologies to support their business and project activities. The reason being payback from investment in such technologies can extend beyond a 12-month period. Consequently, the money invested for initial set-up becomes dead investment for this period. Most SMEs are unable to sustain this investment.

The barriers to the effective use of e-business can be overcome if the infrastructure for e-business use is properly created. Security issues can be handled through firewalls and secure encryption technologies. Currently, most of the communication, both within and outside construction organizations takes place by exchanging e-mails. Most of the e-mail messages are routed between the Internet service providers over public telephone networks and therefore are no more (or less) secure than the conventional telephone calls. The public trust is more biased towards telephone calls than electronic mails. There are also concerns about the legal admissibility of electronic documents. Governments recognize the need for clarity in e-legal matters and measures to address this issue are in place. The Electronic Commerce Bill, for example, states how finger on keyboard will be legally equivalent to pen on paper. This bill also addresses other issues such as trust and reliability issues. One of the measures taken to address trust and reliability issues is the Data Protection Act 1998, which gives consumers the right to object to the use of their personal data for direct marketing.

Some other challenges to e-business adoption in the construction industry have been summarized by Zou and Seo (2006). These include challenges posed by:

- *Information management systems*: In the form of e-mail, Websites, and Internet services, advancement of e-business has generated an enormous wealth of data which leads to information overload. The sharing and transferring of information governs supply chain participant's activities, which serves as a core function of the supply chain. However, due to fragmentation of information from various communication channels, effective logistics of information management to have the purposeful information accessible when required have become laborious and time-consuming activities, and inefficient management of information have reduced the benefits of using e-business. In addition, the potentially enormous data collected from both internal and external communication points involves significant information management load in security, filtering, consistency checking, data cleaning, storing, knowledge discovery, and knowledge integration, which resulted in rather challenging for information management and knowledge integration (Badii and Sharif, 2003).
- *Organizational policies and management*: The introduction of new infrastructure such as e-business systems affects business operations of organizations, and this requires adaptation of a new underlying operation and management philosophy. This change affects core component of organizations both management and employees, such as goals, technology, vision, training, policies, culture, mission, and business strategy.
- *Human resources and culture*: An organization may not possess appropriate skills to manage new innovation technology (such as e-business), which may not be embedded with an underlying supportive culture. Organizational culture contributes a significant part in implementation of innovation that involves different professionals working

together to meet the project objectives and enhance performance, which requires 'no-blame' culture to encourage people to experiment with new concepts. Furthermore, contribution by staff in task execution and management is crucial and their performance can significantly affect the success and failure of the organization (Cheng *et al.*, 2001). The chances of a successful implementation are miniscule in an organization where the current working environment is not ready for changes imposed by e-business (see Chapter 4 for details of organizational readiness for e-business). By training staff resilience and trust in e-business can be considerably improved. Without trust it is less likely that staff will commit to using e-business, thereby jeopardizing it success. Several experts (Hjelt and Bjork, 2006; Ruikar *et al.*, 2006; Zou and Seo, 2006) acknowledge that barriers to e-business are no longer technical or cost-related, but are related to business models and psychology. The main challenges, therefore, lie in addressing the psychological factors of taking the systems (such as e-business) into comprehensive use and overcoming resistance to change. Such psychological barriers can be overcome if organizations engage in proactive measures that focus on creating a positive organizational culture that informs, equips, and encourages staff to learn, adopt and adapt!

2.7 Summary and conclusions

It is clearly evident that the use of e-business in construction can yield several business benefits. There are, however, some challenges and issues that need to be addressed to ensure a successful implementation. Some of the issues are rooted in the nature of the construction industry, which makes project management and communication complex activities. When, effective implemented, e-business can resolve project communication and information dissemination problems. It offers several business benefits to the construction industry in terms of greater collaboration, efficient dissemination of project information, and cost reduction (as the electronic exchange of documents offsets printing, copying and transport costs). The barriers to e-business such as security and privacy issues are being addressed through the introduction of new legislation and data encryption standards. The critical element is to overcome the psychological barrier and getting people to understand and buy into the system. Clearly, the construction industry needs to make the investments necessary to reap the benefits of e-business and overcome the current barriers. This will entail investing in the enabling technologies, exploring new ways of working with the available technologies, and contributing to the development of customized solutions that meet the industry's growing needs.

Currently the larger companies are championing the use of e-business in construction and some of the big players in the construction industry have already realized the benefits of adopting e-business. This is evident from the case studies and examples, discussed in Chapter 3. Wider dissemination of the business benefits and potential risks can improve user confidence and increase the possibility of a successful implementation.

References

Badii, A. and Sharif, A. (2003) Information management and knowledge integration for enterprise innovation. *Journal of Logistics Information Management*, **16**(2), 145–155.

Chaffey, D. (2002) E-business and E-commerce Management, Prentice Hall, Harlow.

Cheng, E.W.L., Li, H., Love, P.E.D. and Irani, Z. (2001) An e-business model to support supply chain activities in construction. *Logistics Information*, **14**(1/2), 68–78.

Clinton, W. and Gore, A. (1997) *A Framework for Global Electronic Commerce*. Washington DC (available at: http://www.iitf.doc.gov/eleccomm/ecomm.htm).

Damanpour, F. (2001) E-business e-commerce evolution: Perspective and strategy. *Managerial Finance*, **27**(7) 16–33(18) Emerald Group Publishing Limited.

Elliman, T. and Orange, G. (2000). Electronic commerce to support construction design and supply chain management. *International Journal of Physical Distribution and Logistics Management*, **30**(3/4), 345–360.

Hejlt, M. and Bjork, B. (2006) Experiences of EDM Usage in Construction Projects. *ITcon*, **11**(Special issue on e-commerce in construction) (K. Ruikar, guest editor), pp. 113–125 (http://www.itcon.org/2006/9).

Kalakota, R., and Whinston, A. B. (1997) Electronic Commerce – A Manager's Guide. Reading, Massachusett: Addison-Wessley Publishing Company Inc., USA.

Laudon, K.C. and Laudon, J.P. (2002) *Management Information Systems*, 7th edition. Prentice-Hall, New Jersey and London.

Learnthat, (2004) Available at: http://www.learnthat.com.

Lewis, T. (1999) *Electronic Data Interchange in the Construction Industry: Volume 1*. PhD Thesis, Department of Civil and Building Engineering, Loughborough University.

Needleman, R. (2000) *Build to Suit*. Available at: http://www.redherring.com/cod/2000/0605.html.

OECD, (1999) *The Economic and Social Impact of Electronic Commerce: Preliminary Findings and Research Agenda*, Organisation for Economic Co-operation and Development, Online Bookshop, ISBN: 9264169725, February.

Ruikar, K., Anumba, C.J. and Carrillo, P.M. (2006) VERDICT – An e-readiness assessment application for construction companies. *Automation in Construction*, **15**(1), 98–110.

Thorbjornsen, T. and Descamps, C. (1997) *e-Commerce – Barriers and Opportunities*. Available at: http://www.infowin.org/ACTS/IENM/NEWSCLIPS/arch1997/971193no.html.

Timmers, P. (2000) 'RE: Consumer – Administration category of e-commerce'. E-mail to K. Ruikar, K.Ruikar@lboro.ac.uk, from Paul.Timmers@cec.eu.int, dated 3 November 2000.

Ugwu, O.O., Anumba, C.J. and Kamara, J.M. (2000) Integration of customer requirements with products and services on the Internet. In *Proceedings of the UK National Conference on Objects and Integration for Architecture, Engineering and Construction*. Watford 13–14 March 2000.

Unysis (2004) Available at: http://www.unysis.com.

Zou, P. and Seo, Y. (2006) Effective applications of e-commerce technologies in construction supply chain: Current practice and future improvement. *ITcon*, **11**(Special issue on e-commerce in construction) (K. Ruikar, guest editor), 127–147 (http://www.itcon.org/2006/10).

3 e-Business: The Construction Context

Kirti Ruikar, Chimay J. Anumba and Patricia Carrillo

3.1 Introduction

The level of uptake of e-business in construction varies from company-to-company and depends on a range of factors such as the type of the company, its role within the construction supply chain and the level of information and communication technology (ICT) competence, amongst others. The progression towards a 'holistic' approach to e-business adoption is gradual as companies develop competence and confidence in the use of new and emerging technologies like e-business. The early years, for example, have seen the Internet being used as a vehicle to promote companies by the dissemination of company product and service information with the view of increasing company profile and subsequently the product sales through online promotion. Other trends from the late 1990s to early 2000 include the use of e-business solutions such as e-procurement through Web directories and search engines, where construction-specific Web directories and search engines enable users to access wide-ranging information about the construction industry including information about construction product suppliers, stakeholders and jobs. With time and maturing experience there has been a steady increase in the use of more sophisticated e-business solutions. This is especially true among the more progressive organizations, which use specialized tools such as project extranet applications and e-tendering tools. This is a significant step forward from the traditional, sometimes arduous paper-based processes. e-Tendering, for example, facilitates the entire tendering process from advertising of the requirement through to placing the contracts. Clearly, e-business can yield business benefits to construction end-users and overcome some of the inherent problems of traditional methods.

Strong information technology (IT) capabilities have been a competitive necessity in nearly every industry sector. The post- Latham (1994) and Egan (1998) era has seen many construction firms investing in technology tools to improve business performance, which subsequently led to an increase in technology investments in construction firms. Many firms, however, fail to recognize that simply investing in technology does not guarantee successful implementation. Adoption of any technology for achieving business targets requires major changes in an organization, its current practices, systems, processes and culture (people). These need to be coupled with correct management practices and strategies. Recent studies have shown that 'good management is good business' and firms are more likely to yield business benefits by marrying good management practices with IT investments (Appel *et al.*, 2004) than by simply increasing their IT spending.

The trends and attitudes towards e-business adoption have always been mixed. On the one hand are the enthusiastic advocates, while on the other the doom-laden opponents who have yet to accept and consider the Internet as an alternative to the traditional tried and tested methods of carrying out business. Part of the scepticism stems from the uncertainty and attitudes towards change, the main challenge, therefore, lies in addressing the psychological factors of taking the systems into comprehensive use and overcoming resistance to change.

This chapter discusses e-business in the construction context including the development of e-business in construction and some of the issues associated with its adoption. It also discusses the key findings of an industry-wide survey aimed at identifying trends and attitudes towards e-business adoption in the UK construction context. The development of a re-engineered business process that highlights opportunities for use of innovative e-business tools within the construction process is also discussed. Finally, this chapter includes findings and analysis of industry-specific case studies that were conducted to assess the impact of specific e-business applications on their end-user processes.

3.1.1 e-Business adoption: An industry perspective

An industry-wide survey conducted in late 2001/early 2002 (Ruikar *et al.*, 2002) attempted to identify the use of IT and e-business within the UK construction sector with the aim of identifying barriers, enablers and the potential of using e-business. It helped in identifying, examining and analysing the industry-perspective on e-business adoption and helped in developing recommendations for the effective uptake of e-business within the construction industry. The survey findings were critical in establishing the readiness of UK construction industry to adopt IT and e-business technologies. It also helped in identifying the barriers and enablers to the implementation of the technologies in the routine construction business processes.

The survey questionnaire was distributed by post to a random sample of 145 construction organizations encompassing various construction disciplines including architects, engineers, contractors, manufacturers and suppliers within the United Kingdom. The overall response rate for the survey was 22%. The survey results, in early 2002, indicated that:

- There was a considerable usage of IT applications in the surveyed construction organisations.
- The level of IT investments largely depended on the size of the organization (i.e. larger organizations had higher budgets).

- e-Business tools were still in the early stages of implementation in most construction organisations.
- There were very few performance measurement tools available to quantify the exact benefits of e-business.
- Most respondents were unsure of the precise benefits of e-business to their respective organizations and to themselves in particular.
- Issues related to Internet security and lack of standards for information exchange across networks, were the two main barriers for using e-business.
- Although security issues were regarded as top priority at a cross-disciplinary level, these were not considered high priority in IT implementation, which is usually within the organization itself.
- Cultural issues, associated with the transition from traditional methods of working to the use of new tools, were seen as a major barrier.
- Issues associated with using the Internet, such as the invasion of privacy and unsolicited mail were not seen as major deterrents for e-business adoption.

In 2001–2002, when the survey was conducted, it was a common view among industry practitioners that the objectives for using e-business technologies in construction were not clearly defined, primarily due to a lack of a well-defined business process model that integrated e-business with the infrastructure and legacy systems of construction companies. Also, for adopting e-business into the day-to-day working of construction projects, companies would *have to* radically alter their traditional processes of managing construction projects and also the ways in which project partners collaborate and communicate with one another. This survey recommended that:

- There was a need to develop business strategies to adopt e-business, including the most appropriate e-business business model(s) for the construction industry.
- Construction organizations needed to explore the new opportunities offered by e-business and re-engineer their business process to maximise benefits.
- Changes in the construction business process due to the adoption of e-business needed to be continually monitored and documented to develop best-practice strategies.
- More e-business performance measurement tools needed to be developed as technology usage matures.

3.2 e-Business and the construction business processes

The survey findings showed that in the early years of its adoption, the exact benefits of e-business in construction were not clear. The reality of the situation, however, was that in other industries e-business models had been considerably successful and the resulting automated processes were faster and more efficient than their traditional counterparts. The most successful example is that of Amazon.com in the retail industry. These success stories made a compelling case for the need to document the effects of incorporating e-business applications into construction business processes and demonstrate opportunities for e-business in construction. By developing a business process model using the principles of business process re-engineering (BPR) it was possible to illustrate how, with the use of innovative e-business applications, different members of the construction supply chain could derive business benefits.

3.2.1 BPR for e-business

Continuous demand for better performance and quality of services and products from customers and clients has forced industries to continually improve their work methods and hence their business processes, which are, at a basic level, a set of activities that transform inputs into outputs for another person or process using people and tools (ProSci, 2001). The process of changing business processes is an iterative one (Greenberg, 1996), in which:

- Current work methods are documented.
- Customer/client expectations are recognized as a means to measure the process effectiveness.
- The process is performed and the process results measured against set standards.
- Improvement opportunities in the process are identified based on the data that has been collected.
- Improved process is implemented and its effectiveness against the set standards is measured.

These iterations can be repeated in cycles to achieve continuous process improvement, also known as business process improvement or functional process improvement. The resulting outcome is a gradual and incremental improvement of business processes.

The advent of the Internet has impacted individuals and businesses on several levels demanding changes in relatively short time spans often necessitating dramatic business process improvements rather than incremental changes. Furthermore, the pace of growth of modern day technologies have forced companies to rapidly bring in new resources and capabilities into their businesses in order to raise the competitive bar through responsive and dramatic improvements to their business processes. BPR, is one such approach that can lead to rapid changes and dramatic improvements.

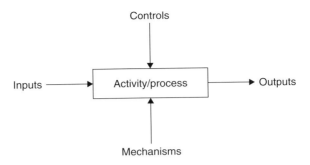

Figure 3.1 IDEF0 notation

BPR is 'the fundamental rethinking and radical redesign of business processes to bring about dramatic improvements in critical, contemporary measures of performance, such as cost, quality, service, and speed' (Hammer and Champy, 1993). BPR can be facilitated through the use of business process mapping methods such as IDEF0 methodology. IDEF0 is an acronym for Integration DEFinition language 0, a function modelling language that is based on SADT (structured analysis and design technique) developed by Douglas T. Ross and SofTech, Inc. (Ward, 2001). Application of IDEF0 method of process modelling to a system such as the construction business process results in a hierarchical series of diagrams, text and glossary cross-referenced to each other. The two primary modelling components are functions (represented on a diagram by boxes) and the data and objects that inter-relate those functions (represented by arrows). The diagrammatic representation of IDEF0 methodology can be seen Figure 3.1.

3.2.2 A representative BPR model

To demonstrate the potential of using e-business in construction a representative BRR model was developed using the information channel (IC), one of the UK's leading collaboration tools at the time (Construction Plus, 2001). This representative business process model provides an insight into how the then current working practices of the construction supply chain could be better managed using the IC. To achieve this an assessment of the IC for its functions and capability was carried out. An initially functional level analysis showed that the IC creates a hub-centric network of information and communication between project partners and uses the principle of a single source for information sharing, thereby reducing the number of communication passages. Thus, for a given project, if any one of the stakeholders updates the project information, then all other relevant stakeholders are immediately informed of the revisions. Such functionality allows fast and easy interdisciplinary communication in a secure environment through the central hub.

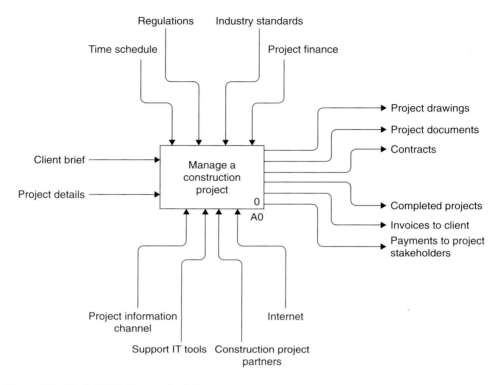

Figure 3.2 Node IC/A0: The context diagram

At the time this BPR model was developed, it was proposed that the IC would incorporate an additional feature called the I-components (intelligent components) that would facilitate the capture of 'intelligent' data during the project lifecycle. It was projected that the I-components would provide the same advantages as the IC, but in relation to the content of the documents. Being intelligent components, the I-components could learn about themselves as the project progresses. They could also be programmed to know what they are, where they are and how they should respond to their location in space, functional area or perhaps what they are connected to, for example, a door. It was also foreseen that the I-components could support the use of construction modelling systems by allowing users to exchange design components for the manufactured equivalents that contractors actually purchase – and therefore test for any fits or clashes (BIW, 2000). The re-engineered construction product procurement process incorporates the use of these I-components. The model presented is decomposed into the following six levels:

(1) Node IC/A0: The context diagram (Figure 3.2) representing the top-level process giving a generic view of managing construction projects.

(2) Node A0: Represents the process of managing project drawings.
(3) Node A2: Represents a specific process of managing project architectural drawings.
(4) Node A25: Proposes the method of finalizing architectural drawings.
(5) Node A253: Represents the selection process of a product supplier (in this case the door supplier as demonstrated in Figure 3.2).
(6) Node A2535: Describes a specific product ordering process (i.e. door ordering process) using the IC.

Node A253 product supplier selection process

The example demonstrated here illustrates opportunities for using e-business in construction processes. The example Node A253 described is that of an innovative procurement process. Node A253 incorporates the concept of using I-components for the door ordering process. The details of all other nodes detailed above can be found in Ruikar *et al.* (2003).

The process of selecting a product supplier (in this case the door supplier) requires input from the client brief, door detail drawings and documents, such as door specifications (see Figure 3.3). Typically, the first step involves accessing the door supplier's database using the IC. This database is an online interactive database of the door supplier's services and all product (i.e. door) information. The suppliers can maintain their product and service data online to ensure that only accurate and most up-to-date information is available. Using the IC search engine, the user can search for relevant door information that meets the required product specifications. For example, a search for a specific door type will list the door suppliers whose products match the door specified. The user can use these search results to compare the different door suppliers using criteria such as cost, quality, availability, delivery time, etc. Following the comparison process the user can shortlist door suppliers. Using the IC, the user can view online catalogues of shortlisted door supplier's Web pages and issue tenders by invitation to these door suppliers. Once the tenders are analysed, the door supplier can then be selected. This entire process, being electronic, can be documented to form a part of the project audit trail document.

Suppliers who advertise products on the IC can update product information regularly. Typically, this information may include product specifications that are defined by the product class, the cost of the product, its availability and quality assurance (see Figure 3.4).

For a given product, the typical classification can include product dimensions, finishes, physical properties, its chemical composition and its use. The information stored on such a system will not only be of use to the company that supplies the product, but also the end-user who can access accurate, updated product information as and when required. For example, the quantity surveyor (QS) can have a better chance of producing an

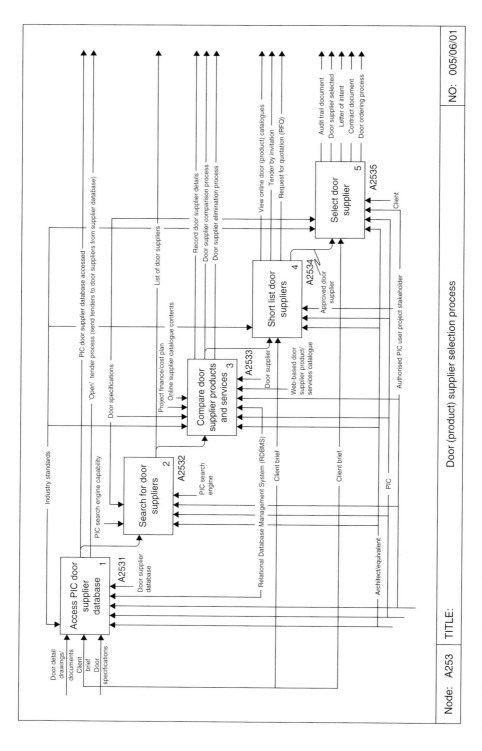

Figure 3.3 Door supplier selection process

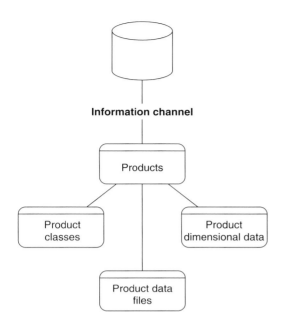

Figure 3.4 Supplier products database

accurate schedule of elements that are required in the construction project itself and he/she can obtain input for the bill of quantities straight from the supplier's database. The door (product) suppliers would benefit from such electronic business transactions as it helps maintain a record of door orders and thus enables better management of product inventory. Furthermore, product sales can be improved by monitoring and coordinating the sales inventory and financial data. Such data can help the management with a complete picture of the company's operations on a day-to-day basis. Suppliers can also access product and order information to establish sales figures and product popularity in terms of which product appeals to the users and why?

The re-engineered business process described here highlights opportunities for using innovative e-business tools within the construction process and the possible business benefits to different stakeholder groups within the supply chain. In order to fully realize the potential of e-business, however, construction companies would have to radically alter traditional practices, especially the ways in which projects are managed and project partners collaborate and communicate with each other. Such changes need to be continually monitored and disseminated to inform late adopters of the impact of specific e-business applications on their end-user business processes.

3.3 e-Business applications and end-user construction companies

Following the examples of other industry sectors, a small, but increasing number of construction organizations are beginning to adopt e-business for performance improvements and addressing the adversarial inter-organizational relationships and fragmented processes. In the past few years, the UK construction industry has witnessed the emergence of a number of e-business applications to monitor, control, manipulate and store project information and to make it available to all project participants across the construction supply chain. This would lead to a step change in current working practices. To facilitate this step change it was vital to examine (and document) the impact of specific e-business applications on their end-user business processes. Case studies were therefore conducted to assess the impact of specific e-business applications on their end-user business processes (Ruikar *et al.*, 2005). Two construction-specific e-business applications, described as Products A and B for anonymity, were selected for case studies. These case studies mainly focused on three key aspects:

(1) Investigating the integration of specific e-business applications into the business processes of their end-user companies.
(2) Examining the impact of these applications on the traditional business processes of the construction end-users and their supply chains.
(3) Establishing the effectiveness and suitability of these applications in terms of their benefits and barriers.

3.3.1 Product reviews

In the review process, Products A and B were reviewed for their scope and functionality. This provided an insight into the background, scope, functionality and other details of each application and helped to capture key company (ASP) and product-specific information. The next section gives a brief overview of Products A and B and illustrates with an example, the impact of these on traditional construction processes.

Product A overview

Product A is an online project collaboration tool with the help of which construction industry professionals can collaborate with other project partners using Web technologies during construction projects. It enables online exchange of information and documents between construction project partners, including concept drawings, specifications, feasibility studies, detailed design drawings, fabrication drawings, structural design drawings, Operations and Maintenance (O & M) documents and Health

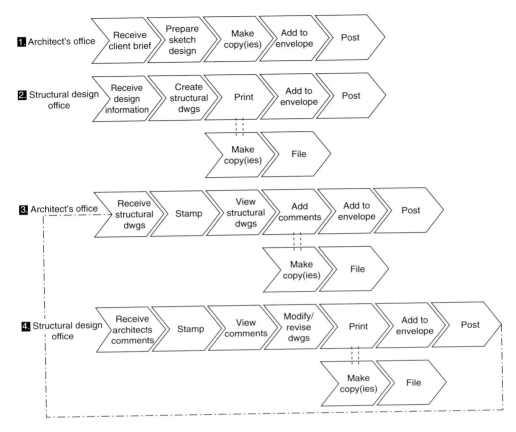

Figure 3.5 Traditional processes

and Safety (H & S) records, among others. Product A generates a permanent database of all project information and documents including drawings, revisions, comments made to prompt such revisions, documents, meeting minutes, progress on site, site photographs and all other project-related data. Using Product A, members of the construction supply chain can communicate and archive information (records of what was done, when, by whom, etc.) throughout the lifecycle of a construction project.

Impact of Product A on the traditional process
This section examines the impact of Product A on traditional construction processes. It does so through an example of the typical processes between an architect's office and the structural design office (see Figure 3.5).

For a given construction project, typically the architect prepares sketch design drawings using client's brief, project specifications and other relevant documents. Paper copies of these documents are filed in the architect's project file and also posted to the structural design office. On receipt of the design information (e.g. design drawings and related documents) from the architects office, the structural engineer prepares structural design drawings, prints copies of these, files a completed drawing set for record and posts the revised/new set to the architect's office for comments/approval. These structural design drawings are stamped on receipt and then viewed. If the structural drawings are satisfactory they are accepted and the process can move to the next level (i.e. the preparation of detail drawings). If, however, comments are added and changes suggested, then copies of the red-lined documents are maintained for record purposes and also sent to the structural design office for corrections or modifications. This process is iterative and is repeated till the design is finalized. Similar iterations are involved in the processes between the structural design office and steelwork fabricator. The resulting process is, thus, complex and time-consuming. Also, reliance on third party services (such as postal, courier services, etc.) results in delays and affects the overall project cost, timescale and budget. This kind of one-to-one correspondence between different project stakeholders makes the project communication network very complex and protracted.

Using Product A the traditional process can be cut down considerably. Project drawings and documents can be uploaded onto the main server, where a permanent/secure database of project information is maintained. Relevant construction disciplines can then be automatically (and instantly) notified and invited to online discussions, commenting online or responding to comments made by others. The resulting process (as demonstrated in Figure 3.6), while being much more efficient, can provide financial and time benefits to its end-users. A study carried out by an international construction management and consulting firm indicated that using Product A their firm had achieved approximately 2% savings in printing and postage costs for a £5 million construction project.

Product B overview

Product B is an Internet-based supply chain solution for trading and material procurement. Trading processes such as sourcing, procurement and administration of plants and materials are automated. Using Product B, documents such as purchase orders, invoices and despatch notes, can be exchanged online between different supply chain partner applications within a secure environment. It provides back office integration with bespoke applications used by different members of the construction supply chain. Also, different applications using different document formats use the central database to send and receive trading documents to/from other end-user disciplines. The central database converts the data into the data formats of the sender or receiver, to enable seamless document exchange across end-user disciplines.

e-Business: The Construction Context   35

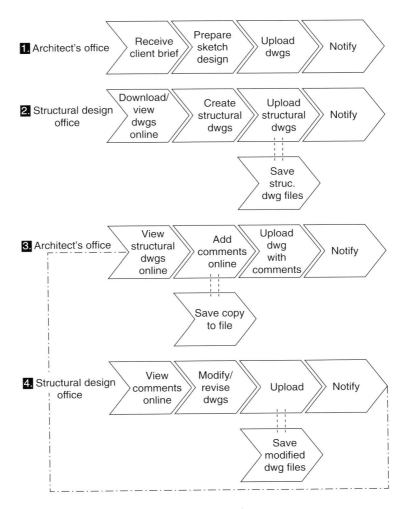

Figure 3.6   Modified processes using Product A

Impact of Product B on the traditional process
A typical materials ordering cycle using traditional methods, involves the main contractor sending out a paper-based Purchase Order to the materials supplier. On receipt of the Purchase Order, the materials supplier acknowledges the order, creates an invoice manually and posts a printed copy of the paper invoice to the main contractor. On receipt of the invoice, the main contractor would stamp a date of receipt. The invoice would be then checked against the Purchase Order details and goods received. If the contractor is satisfied with the goods received a payment is made and the invoice filed away. This traditional trading cycle involves several intermediary processes or *layers*. Figure 3.7 illustrates the various processes involved at the sender and receiver-end of the trading cycle.

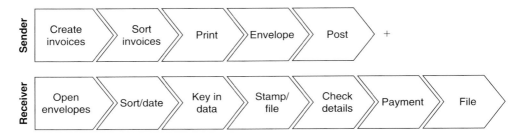

Figure 3.7 Traditional construction trading process

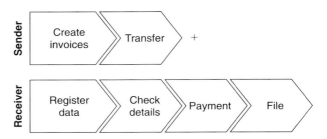

Figure 3.8 Trading process using Product B

Using Product B, the traditional trading process can be considerably simplified either by removal, modification or substitution of intermediary processes with new processes. Using this application the sender can create an electronic invoice and transfer it via the Internet to the receiver. Given the electronic nature of the invoice, no time is wasted in re-keying data or filing the invoice. This simplified trading process also cuts out the need to print and post paper copies of the trading documents. The resulting process, as illustrated in Figure 3.8, has benefits in terms of time, money and efficiency. A survey carried out by a leading finance corporation to measure the cost benefits of using Product B showed savings of over 60% on invoice processing. This same survey also showed that while the traditional construction trading cycle takes eight days (from creating invoices to making final payments and filing), using Product B it takes four days for the same trading cycle. Thus a reduction in the cycle time by half and an overall cost reduction of more than half that of the traditional cost can be achieved.

3.3.2 *End-user case studies*

With a view to capture end-user processes 'before' and 'after' using e-business applications, the drivers for e-business adoption, reasons for companies to engage in e-business, the skills and competence requirements, and the business benefits and limitations of e-business applications, case studies with end-users of Products A and B were conducted. Management level staff, such as senior project managers and IT managers, from each end-user company were interviewed using semi-structured telephone interviews. In all eight end-user companies were interviewed. These include four contracting, one quantity surveying and project management firm, one architecture and interior design firm, one construction management consultancy and one materials supplier. All, except the supplier firm were large companies employing in excess of 500 employees.

From the case studies it was clear that end-user companies had accrued several business benefits from using e-business applications. Several factors can influence the adoption of technology within construction organizations. Technology adoption can be management driven, client driven, market driven or project driven. According to marketing guru, Geoffrey Moore (2003), technology adoption within an industry sector is based on a traditional Technology Adoption Life Cycle, represented by a bell curve. This bell curve comprises of industry end-users, progressing from innovators, early adopters, early majority, late majority, to laggards. While innovators are 'technology enthusiasts', early adopters are visionaries who have the insight to match an emerging technology to a strategic opportunity, and are driven by a 'dream'. Their core dream is a business goal, not a technology goal. The early majority are pragmatists who share some of the early adopters' ability to relate to technology. However, they require well-established references, before they invest in such technologies. The late majority have a conservative approach to technology and are pessimistic about the ability to gain value from technology investments and start using it only under duress. Laggards are those who simply do not believe in technology.

This analogy can be used for the adoption of e-business within the construction sector. Case study findings indicate that the interviewed end-user companies fall within the early adopters' category of the adoption curve as shown in Figure 3.9. In almost all end-user companies the use of e-business was driven by the company's senior management. These companies believe that e-business is the way forward for project management within the construction industry and have adapted the technology to match their business needs. Quoting one response:

> 'use of innovative technologies such as this, has been a primary differentiator that sets us apart from the rest and reflects our company's drive towards innovation. This is because we fundamentally believe that that's the way the industry is going to go and we would like to be at the forefront of innovation.'

One large construction management company, for example, made a strategic decision to implement Product A. It had successfully used EDM (electronic document management) tools in the past to manage large

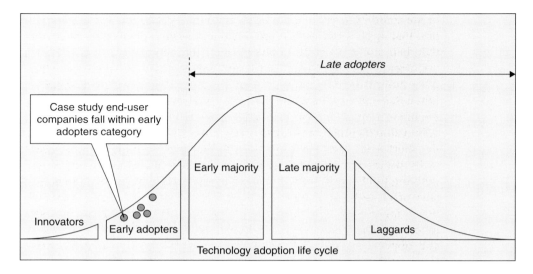

Figure 3.9 Technology adoption life cycle. (*Source*: Adapted from Moore, 2003)

construction projects. The company has a policy for continual improvement and innovation, hence it was only natural to migrate to the next level that is strategic move to use Web-based EDM systems such as those facilitated by e-business. Another company, a large contractor and property development company, believes in early technology adoption and it is a company policy to always be at the cutting edge of 'whatever' (e.g. technology). For others such as a small construction materials supplier, using e-business was primarily project led with a push from the prime contractors. In hindsight, however, the supplier acknowledges that their competence in using e-business has opened new opportunities, which were not previously possible. This has given a competitive edge, which can be beneficial in future projects where clients seek expertise in this specific area. These business benefits as demonstrated by 'early adopters', act as drivers for wider adoption of e-business amongst the 'late adopters' of the construction industry.

Early buy-in from senior management is a crucial factor influencing successful adoption. This can be 'a hard nut to crack', but it is not an impossible task. More buy-in can be encouraged through previous examples of success stories. Geoffrey Moore (2003), a management guru, says that 'part of what defines a technology market is the tendency of its members to reference each other when making buying decisions is absolutely key to its success'. Quite often the scepticism and techno-phobia, especially within the early majority, stems from the probable risks. Risks most companies are not willing to take. In addition, cultural issues associated with technology adoption are considered a key factor inhibiting wider use of e-business in construction. Software tools that 'do too much' or are complex to use, are not easily adopted by the industry. It is important that application service providers and system developers recognize that their tool is simply an enabler in a bigger process.

Currently, there is evidence of process improvement resulting from automation of processes and removal of bottlenecks leading to rationalization of procedures. There is, however, no evidence of process re-engineering or development of any new processes that have led to paradigm shifts within the construction process. Such practices may work in the short term, but have limitations in that the end-users are not necessarily utilizing the technology to its full potential and are therefore not deriving full benefits from it. Instead of regarding e-business as an extension of IT and fitting it into the existing business processes, construction companies need to recognize that it is a radically different approach to conducting business and therefore should explore new processes and opportunities, which may only be possible because of e-business. Changes that occur in the construction business process due to the adoption of e-commerce need to be continually monitored and documented so that a best-practice strategy for e-business adoption in construction can be formulated.

Currently, application service providers develop applications that support individual processes (e.g. design management, procurement, trading, etc.). Thus, the resulting processes, although automated, are still fragmented. Existing e-business solutions have the capability of providing a 'single' solution that encompasses the project lifecycle – from inception to completion including the facilities management phase, thus in effect overcoming the fragmentation issue. The industry, however, is still not ready to accept the radical changes that are essential to realize the full potential of e-business implementation. At present, the industry has accepted and adopted e-business applications that have minimal impact on their processes and hence are easy to implement and use. Such tools facilitate the construction process and increase the probability that the project will meet its target cost and time schedule. In this respect, e-business contributes towards improving project performance. To gain maximum benefits from adopting e-business, however, the correct strategies and implementation plans have to be developed, communicated, implemented and monitored. Taking this into account, construction companies who are currently using and those who have yet to use, e-business need to take measures to successfully adopt, use and benefit from the technology. It is important for companies that seek to adopt e-business tools to assess their 'e-readiness' for adopting e-business to ensure a productive and beneficial implementation. To address this need an e-readiness model for construction organizations and a prototype application, VERDICT that assesses e-readiness of the construction sector was developed and implemented. This e-readiness model and prototype application has been discussed in detail in Chapter 4.

3.4 Summary

This chapter discussed e-business in the construction context including the developments of e-business in construction and some of the issues associated with its adoption. It presented the key findings of an industry-wide e-business survey and the development of a re-engineered business process that highlights opportunities for use of innovative e-business tools within the construction process. Finally, this chapter discussed the analysis of industry-specific case studies that were conducted to assess the impact of specific e-business applications on their end-user processes. It is evident that the construction industry is ripe for the adoption of e-business, as it stands to reap numerous benefits. These benefits are highlighted in various chapters of this book.

References

Appel, A.M., Dhadwal, A., Dorgan, S.J. and Dowdy, J. (2004) When IT creates value. *McKinsey on IT* (Winter).

BIW (2000) *Project Collaboration Overview*, previously Building Information Warehouse now BIW Technologies, UK.

Construction Plus (2001) *Internet Business of the Year: Managing Construction Projects Online*. Construction Plus, London.

Egan, J. Sir (1998) *Rethinking Construction: Report of the Construction Task Force on the Scope for Improvement the Quality and the Efficiency of UK Construction*. Department of the Environment, Transport and Regions, London.

Greenberg, L.J. (1996) *Business Process Reengineering: Constantly Adapting To Change*. Available at: http://www.earthrenewal.org/bpr.htm (accessed 29th April 2002).

Hammer, M. and Champy, J. (1993) *Reengineering the Corporation, A Manifesto for Business Revolution*. Nicholas Brealey Publishing, London.

Latham, M. Sir (1994) *Constructing the Team: Final Report of the Government/Industry Review of Procurement and Contractual Arrangements in the UK Construction Industry*. HMSO, London.

Moore, G.A. (2003) *Inside the Tornado: Marketing Strategies from Silicon Valley's Cutting Edge*. Capstone Publishing Ltd., West Sussex, UK.

ProSci, (2001) *Reengineering Tutorial Series: Introduction to BPR*, BPR Online Learning Center, Available from: http://www.prosci.com/mod1.htm.

Ruikar, K., Anumba, C.J. and Carrillo, P.M. (2002) Industry perspectives of IT and e-commerce. In: *Proceedings of the 3rd International Conference on Concurrent Engineering in Construction at University of California*, Berkeley, July, 26–40.

Ruikar, K., Anumba, C.J. and Carrillo, P.M. (2003) Reengineering construction business process through electronic commerce. 'Managing quality in e-operations', special issue, *The TQM Magazine* (Emerald Press, UK), **15**(3), 197–212.

Ruikar, K., Anumba, C.J. and Carrillo, P.M. (2005) End-user perspectives on the use of project extranets in construction organisations. *Engineering Construction and Architectural Management* (Emerald Press, UK), **12**(3), 222–235.

Ward, M.A. (2001) *Building Structural Frame: Steelwork – Reading AP Models*. Available at: http://www.leeds.ac.uk/civil/research/cae/step/ap230/ap_revw/rd_modls.htm.

4 Organizational Readiness for e-Business

Kirti Ruikar, Chimay J. Anumba and Patricia Carrillo

4.1 Introduction

With the growing importance of the Internet, companies across several industries, including construction, are increasingly leveraging the Internet to achieve competitive advantage (Cheng *et al.*, 2003). Internet-based tools such as project extranets are being used to manage construction projects. Such tools can be used to monitor, control, manipulate and store project information and to make it available to all participants of the construction supply chain (Alshawi and Ingirige, 2002). Examples of Internet-based tools include a computer-mediated tendering system for services or contracts, purchasing of materials via the Internet by a contractor, project extranets for project management and specifying products online by a manufacturer (ITCBP Intelligence, 2002). All these tools can be categorized under the broader umbrella of e-business for construction as they support and/or facilitate business functions such as trading, exchange of data and information, and automation of the business processes and workflows.

Research studies (Ruikar *et al.*, 2001; Anumba and Ruikar, 2002) and recent industry and research publications (Laudon and Laudon, 2002; Stephenson and Turner, 2003) have documented the possible benefits and business opportunities for companies using e-business tools such as project extranets. In spite of these documented benefits, the UK construction industry has been relatively slow in the uptake of these tools in their day-to-day workings. A survey of the UK construction industry, undertaken by the Construction Products Association (CPA, 2000), predicted that by 2005, 50% of the industry's business activities would be undertaken using e-business. However, another survey carried out a year later by the same organization indicated a considerable reduction in these projected figures to 22% (which is less than half of what was initially predicted), indicative of a much slower uptake than anticipated. The construction industry stepping back from the initial 'dot-com fever' was seen as the main reason of this change (CPA, 2001). Additionally, some other factors that have also contributed to this slow uptake are:

- Since e-business technology is relatively new there is limited availability of information or feedback on the technology's performance on previous construction projects.

- As with most technologies, it can be difficult to gauge the quantitative return on investment (ROI) from using new technologies such as e-business.
- The teething problems and changes in working culture and practices which are required initially, with the adoption of any new technology, very often deters new users.

Although the uptake of e-business in the UK construction industry has been relatively slow (Construction Industry Times, 2002; Stewart and Mohamed, 2003), it can be seen that the construction industry has now realized the enormous potential for its use in the construction sector. The UK construction sector is trying to maximize the use of e-business through several industry and government-backed initiatives [e.g. M4I (Movement for Innovation), CBPP (Construction Best Practice Programme), CPA (Construction Products Association), Construct-IT, etc.] that promote research into the use of emerging technologies, such as e-business, in construction. It is now the industry's view that, e-business is here for the long run and it will not be long before it becomes an industry standard.

Implementation of any new technology such as e-business for achieving business targets requires major changes in an organization, its current practices, systems, processes and workflows (ITCBP Intelligence, 2003). The correct strategies and implementation plans have to be developed, communicated and implemented at all levels. Since this is not easy, issues such as 'buy-in', defining a strategy, selecting a system, developing a training programme, defining operating procedures, modifying organizational structures, reviewing use and extending use need to be thoroughly researched (ITCBP Intelligence, 2003). Taking this into account, construction companies who are currently using and those who have yet to use, e-business tools need to take measures to successfully adopt, use and benefit from e-business. It is important for companies that seek to adopt e-business to analyse their businesses to ensure a productive and beneficial implementation of these tools that is:

- They need to evaluate the impact of using e-business tools on their day-to-day business processes.
- Assess their 'e-readiness' for adopting e-business.

This chapter focuses on the development of an e-readiness assessment tool (VERDICT) for construction organizations. The next section defines e-readiness and describes the adopted methodology. This is followed by a review of readiness assessment models and detailed description of the development and implementation of the VERDICT tool. An example is used to illustrate the operation and features of the system and aspects of system evaluation are also presented.

4.2 Methodology for e-readiness

e-Readiness can mean different things to different people, in different contexts, and for different purposes. The authors define e-readiness is as 'the ability of an organization, department or workgroup to successfully adopt, use and benefit from information and communication technologies (ICTs) such as e-business'. It is important for companies that seek to adopt e-business to undertake an analysis of their businesses to ensure a productive and beneficial implementation of these tools (i.e. they need to evaluate their 'e-readiness'). The approach adopted in the development of an e-readiness assessment tool for construction organizations is presented in this section.

Triangulation methodology that uses both qualitative and quantitative approaches, is adopted for the development of the e-readiness model that assesses the readiness of construction organizations for e-business. Using this method, theories can be developed qualitatively and tested quantitatively. Triangulation increases the validity and reliability of the data, since the strengths of one approach can compensate for the weaknesses of another (Sunyit, 2004).

A systematic two-stage approach has been adopted for assessing e-readiness. The first stage involves developing an assessment model for gauging the e-readiness of construction organizations for using e-business applications. Using a qualitative approach, a review of existing literature is carried out. The best suited models in the context of this research study are then adapted to develop a model that assesses the e-readiness of construction organizations. The existing processes, working methods, procedures and practices in construction organizations are also analysed using qualitative methods such as one-to-one discussions, case studies and interviews. The outcome of this led to the development of a set of questions for assessing the overall e-readiness of construction organizations for adopting and implementing e-business technologies. Further, a quantitative approach is adopted to analyse end-user responses (by calculating cumulative and average scores) and presenting the findings graphically.

The second stage involves development and evaluation of a prototype application. The development of this prototype is an iterative process that uses rapid application development (RAD) methodology of software development. RAD is a concept that facilitates faster development of application software (Webopedia, 2004). RAD is performed iteratively through several stages as illustrated in Figure 4.1.

The e-readiness tool was evaluated using a number of methods including self-evaluation and peer reviews during the development phase and then through industry validation of the final prototype software. Details of the evaluation are described in Section 4.6.4 of this chapter.

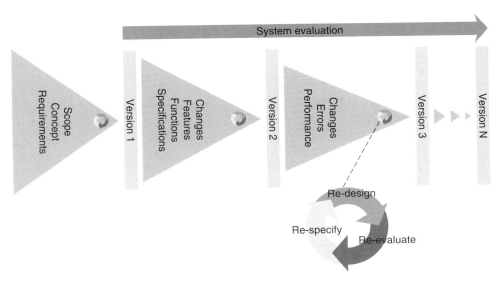

Figure 4.1 Rapid application development using iterative prototyping. (*Source*: Adapted from Maner, 1997)

4.3 Review of readiness assessment models

An increasing number of readiness assessment tools have been developed over the last few years. On the surface, each tool gauges how ready a society or economy is to benefit from information technology and e-business. However, according to Peters (2001) the range of tools use widely varying approaches for readiness assessment, including different methods for measurement. Each assessment tool or model has a different underlying goal and definition of e-readiness. While some gauge the readiness of countries and economies to adopt Internet technologies on a global platform, others are more focused on assessing the readiness of specific industry sectors to adopt Internet technologies.

Several readiness assessment models were reviewed as a part of this study, including those that were not construction-specific. Harvard University's tool called the 'Networked Readiness Index' assesses a country's capacity to make use of its ICT resources. It defines e-readiness as the degree to which a community is prepared to participate in the networked world including its potential to participate in the networked world in the future (Kirkman *et al.*, 2002). On the other hand, APEC's (Asia Pacific Economic Co-operation) E-Commerce Readiness Initiative focuses on government policies for e-commerce (Information Technology and Broadcasting Bureau, 2000; Peters, 2001; Bui *et al.*, 2002). Mosaic's readiness assessment tool aims to measure and analyse the worldwide growth of the Internet (The Mosaic Group, 1998).

While these tools focus on assessing readiness of countries, governments and policies for adopting Internet technologies, some others for example SCALES (Supply Chain Assessment and Lean Evaluation System) assess the readiness to adopt different concepts or approaches for engineering (e.g. readiness assessment tools for concurrent engineering (CE). SCALES was developed for a specific industry sector – the manufacturing industry (K3 Business Technology Group, 2004). It was designed to assess a company's (especially SMEs) readiness for adopting lean manufacturing techniques. RACE, on the other hand, is a readiness assessment tool for CE and is widely used in the software engineering, automotive and electronic industries (CERC, 1998).

Two other readiness models that are of particular relevance are the BEACON model (Khalfan, 2001) and the IQ Net Readiness Scorecard (Cisco Systems, 2004).

(1) The BEACON model: BEACON (**B**enchmarking and **R**eadiness **A**ssessment for Concurrent Engineering in **Con**struction) assesses the readiness of construction companies to improve their project delivery processes through the implementation of CE (Khalfan, 2001). It consists of four elements, which are Process, People, Project and Technology. A commercial software tool has been developed to automate the process of CE readiness assessment for construction organizations. The software takes the user through a series of questions and generates a diagram called the BEACON model diagram that graphically illustrates the assessment results.

(2) IQ Net Readiness Scorecard: This was developed by CISCO and is a Web-based application that assesses an organization's ability to migrate to an Internet business model. It is based on the book Net Ready by Amir Hartman and John Sifonis (2000), which gauges the readiness of IT service providers. The application comprises of a series of statements that fall into four categories – Leadership, Governance, Technologies and Organizational Competencies. Similar to the BEACON model, companies are required to respond to the statements and on completion, they are presented with an IQ Net Readiness Profile.

The model that is described in this chapter combines aspects of these two models and builds on them. The proposed model adopts a similar methodology where the end-users are presented with a set of statements and an assessment of their e-readiness is based on their responses. On completion, the respondents are presented with a report which includes textual and graphical data. Where the proposed model differs from the two described above is that, while the BEACON model focuses on CE and the IQ Net Readiness Scorecard addresses the readiness of technology companies (e.g. software companies, vendors and application service providers (ASPs) to develop applications and profit from what is termed the 'e-conomy', the proposed model assesses the e-readiness of construction organizations to adopt e-business. The readiness assessment tool that is based on the proposed model is called VERDICT (an acronym for Verify End-user e-Readiness using a Diagnostic Tool).

4.4 Verify end-user e-readiness using a diagnostic tool

4.4.1 Background

VERDICT is an Internet-based prototype application that assesses the overall e-readiness of end-user companies and profiles the companies in this regard, based on their responses. The name, 'VERDICT' reflects the overall aim and purpose of the application. VERDICT is developed to aid construction sector end-users to gauge their e-readiness for using e-business technologies such as Web-based collaboration tools. It can be used to assess the e-readiness of construction companies, department(s) within a company or even individual workgroups within a department.

4.4.2 The verdict model

Several research publications (Kern *et al.*, 1998; IBM, 1999; Goolsby, 2001; Basu and Deshpande, 2004) and articles (Emmett, 2002; Fuji Xerox, 2003; Larkin, 2003) indicate that people, processes and technology are the three key aspects that need to be considered for successful implementation of technologies. Emmett (2002) states that together these three elements create business value. However, he further states that 'the people, processes and technology need a leader', just as 'an orchestra needs a conductor'. Emmett draws a parallel with the performance of an orchestra and states:

> 'in an orchestra …. You've got musicians (people), musical scores (process), and musical instruments (technology). But without a conductor, they're more likely to produce noise than music. Even if everyone in the string section plays the right notes at a relatively similar tempo, creating a symphony requires more than following the sheet music.' Therefore '…. An orchestra needs a conductor'.

The same analogy can be applied to the adoption and implementation of new and innovative technologies within construction companies. The 'conductor' in this case is the management. To successfully implement and use any new technology it requires management buy-in and belief in order to plan and drive policies and strategies. The research findings from an industry-wide survey (Ruikar *et al.*, 2002) and case studies (Ruikar *et al.*, 2004; Ruikar *et al.*, 2005), complement this view. The case study findings identified management buy-in and leadership as a critical factor affecting the adoption of technology within construction organizations. The adoption of any new and innovative technology (e.g. e-business, e-commerce) within an organization/department/work-group requires total commitment from the management (or group leader). It is important for the management to buy into the technology to lead the business into successfully implementing and adopting the technology (i.e. the management needs to be e-ready). Thus a fourth category, 'management' is necessary. Taking this into account, the verdict model has been so structured that for an organization to be e-ready it must include:

- *Management* that believes in the technology and takes strategic dynamic measures to drive its adoption, implementation and usage in order to derive business benefits from the technology.
- *Processes* that enable and support the successful adoption of the technology.
- *People* with adequate skills, understanding of, and belief in, the technology.
- *Technology* tools and infrastructure necessary to support the business functions (e.g. processes and people).

All four categories are considered important for an organization to be e-ready. A company cannot be e-ready if it satisfies the requirements of just one category and not the others. For example, even if management, processes and people are e-ready, the fact that the technology infrastructure is inadequate will affect the overall e-readiness of the organization. This example indicates that the company will need to address its technology issues in order to be e-ready. Drawing from the orchestra analogy, *'a memorable symphony performance doesn't happen when the players just assemble with their instruments and scores.'* and, *'the orchestra with the most violinists isn't necessarily going to sound the best.'* Similarly, all four categories – management, processes, people and technology – need to work hand-in-hand and symbiotically (see Figure 4.2).

Many organizations fail to realize that installing a system without first achieving universal buy-in and changing business processes, will result in a software installation, not an implementation of a comprehensive solution to business problems. According to Larkin (2003), if an organization merely completes an installation by automating inefficient processes, it will not realize a long-term positive impact. A successful company-wide rollout includes more than simply buying and installing software. It requires the management to align people, processes and technology to implement a solution that meets business needs. The result is the ability to capitalize on the full potential of the technology investment. Thus, the implementation of new technology needs to be carefully managed and orchestrated. Companies should recognize that in order to successfully

Figure 4.2 Four key elements for an e-ready organization

implement and benefit from new technologies such as e-business, it is essential that the people (who are the ultimate users of the technology) and the process are given due consideration. The technology within the company also needs to be assessed in order to ensure that the company has the necessary infrastructure (ICT infrastructure) to use existing and new or emerging technologies successfully. Further, the company needs clear leadership and direction that is provided by the management in order to successfully implement the technology. These four categories are described in detail below.

Management

Management is an important factor that leads and governs the adoption, implementation and use of technology within organizations through the careful orchestration of business strategies in order to derive definite business benefits. This can be achieved by defining specific business strategies for technology adoption and by ensuring that adequate resources are available in terms of funds, time and man-power.

Management buy-in as an important aspect that can influence the successful implementation and adoption of technology/technologies within a construction organization. Senior managers can authorize investigations into all aspects of current activities to identify areas where improvements can be achieved by changing to new e-business-based systems. However, management should endorse the technology only after investigating the technology for its overall capability and scope. It should

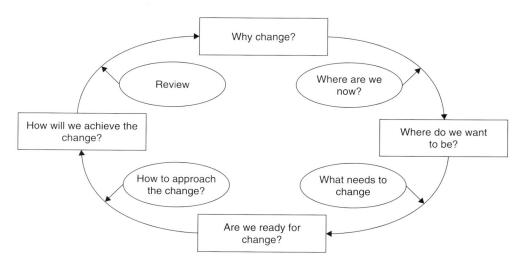

Figure 4.3 Aspects of change management. (*Source*: Adapted from TUV Management and IT Consultancy, 2004)

examine whether the technology has been successfully used in construction or other industries before, including technology reviews (favourable or otherwise). In the absence of such reviews, it is worthwhile identifying and investigating the possible risks and taking adequate measures for minimizing the risks and maximizing the rewards. The adoption of technology will bring about change, and management needs to carefully consider different aspects of how this change will be brought about and managed. These aspects are highlighted in Figure 4.3.

If the management takes into account these aspects of change management, it is more likely than not that the business will be favourably geared towards (or ready for) adopting and implementing e-business. However, it is important that the senior management does not loose sight of its ultimate vision and aim in using the technology (e.g. to derive business benefits). To quote Paul Nitze, a famous American diplomat and strategist, 'One of the most dangerous forms of human error is forgetting what one is trying to achieve' (Hill, 2004). In business terms this can have dire consequences.

Process

Process means a practice, a series of actions, done for a specific purpose (Craig, 2004). It also includes the working rules, ethics and procedures, within and between organizations. It is important to consider the process factor as the adoption of new technology will directly affect an organization's processes and vice-versa. Therefore, companies will need to ensure that the new technology either complements their existing processes or that the existing processes are flexible enough to accommodate the technology. In order to maximize the benefits from technology adoption (e.g. increased transparency, reduced response time and improved integration of activities across the supply chain), organizations need to examine and map their existing processes. This will help in identifying the bottlenecks and devising measures to remove such bottlenecks or process inefficiencies. The process-related change that technology can bring about is fourfold (Laudon and Laudon, 2002):

(1) *Process automation*, which is the most common form of change where organizations use computers to speed up the performance and efficiency of existing tasks and functions.
(2) *Rationalization of procedures*, which involves the streamlining of standard operating procedures, eliminating obvious bottlenecks, so that automation makes operating procedures more efficient.
(3) *Business process reengineering*, which is a powerful form of organizational change in which business processes are analysed, simplified and redesigned.
(4) *Paradigm shift*, which involves rethinking the nature of the business and the nature of the organization itself.

Process change of any nature carries its own rewards and risks (Laudon and Laudon, 2002). Process automation and rationalization of procedures are relatively slow-moving and slow-changing business strategies that present modest returns, involving lower risks. The much faster and more comprehensive change is brought about by process reengineering and paradigm shifts carry higher rewards and risks.

People

The people factor accounts for the social and cultural aspects related to the people within an organization. It takes into account the attitudes, outlook, and feelings of staff within an organization towards change brought about by technology adoption. People make organizations and are important to its success. No matter how carefully the management has geared the business to successfully adopt new technology, it is less likely to succeed to its full potential, if the people are not ready. The people, who are the ultimate users of the technology, need to have the appropriate skills and competency, functional expertise, the right attitudes, a positive mindset and the culture to adapt and adopt.

The people factor is important and can affect an organization's overall e-readiness, because the introduction of any new technology (hence change) will affect the workforce within that organization. It is therefore necessary to assess the organizational culture and the readiness of company staff (people) in accepting new and innovative technologies such as e-business. It is also important to ascertain whether the organizational structure provides an appropriate environment for e-business adoption and use. According to Ostroff (1999), the horizontal organization is well suited for the information age. Such horizontal structures allow for greater flexibility in dealing with today's competitive and rapidly changing business environment. Through the use of e-business tools, projects can be managed in an open environment with more transparency between different members of the team. For an organization in which such an open culture already exists, the likelihood of a 'culture shock' is reduced and therefore the change is less likely to be met with any resistance. However, for those organizations in which there traditionally exists a culture of secrecy and privacy, the people factor may be more of an issue that needs addressing.

Technology

The final category to consider is technology. The technology factor covers all aspects related to IT (information technology) and communications technologies (e.g. Internet technology), which include both the hardware and software usage and its availability within a company, department or workgroup. Also important are the aspects related to the performance of the technology – thus, even if the technology infrastructure is adequate and available, it is of little or no consequence, if it cannot efficiently perform the required functions. For example, an end-user company may have a computer linked to the Internet, but still cannot send large files (e.g. CAD drawings) because the system is not equipped to handle such tasks. The problem in this case is not just confined to that individual company. This is mainly because, technologies such as e-business allow project teams to communicate and exchange data in a collaborative electronic environment. Thus, even if one company in the chain is ill-equipped, it has adverse effects on the entire chain. Given that 'a chain is only as strong as its weakest link', it is an important technology issue that needs to be considered for assessing e-readiness.

Technology is capable of coordinating different activities within and across organizations and also across industries (Laudon and Laudon, 2002). With the help of technology, companies can reduce transaction and document processing costs and time (Ruikar et al., 2004). Processes can be made more efficient and streamlined by removing the obvious bottlenecks. All these benefits, however, cannot be realized if an adequate technology infrastructure is not in place and available to the people in an organization, who are the ultimate end-users. Organizations that aim to use e-business should, as a minimum, be equipped with the basic infrastructure necessary for its operation.

Since e-business enables the seamless and electronic exchange of knowledge, sharing and editing of documents, revision of reports or publications, within or between workgroups any previous experience in using collaborative tools (e.g. electronic document management (EDM), Groupware) is useful. This is especially vital, given that the findings of the case studies suggest that for companies with experience in using applications such as EDM and Groupware, it is easier to migrate to the next level that is Web-based collaboration tools (Ruikar *et al.*, 2005).

4.4.3 Verdict implementation

The VERDICT application consists of a series of judgement statements which fall into the four categories of management, process, people and technology. The end-users may either agree or disagree with these statements, to varying degrees. Verdict relies on the judgement of the respondent (i.e. end-user) as to whether or not he/she agrees with the statement/s in the context of their organization, department or group. The respondent(s) need to ensure that their responses are consistent with their assumptions for example if the responses are in the context of the department (and not the organization), then that assumption must be consistently reflected throughout. The extent to which the respondent agrees or disagrees with the statement is graded on a scale of 1–5, where 1 = Strongly Disagree, 2 = Disagree, 3 = Neutral, 4 = Agree and 5 = Strongly Agree. A 'don't know' option is also included (where don't know = 0 score). The questions/statements are so orchestrated that a response of strongly agree will generate the highest score of 5 points. An average score is calculated for each category. The higher the average score the more likely it is that the end-user company is 'e-ready'. Respondents are required to answer all questions for a meaningful outcome. Once all the questions are completed the end-users are presented with a final verdict with respect to their e-readiness in the form of reports, which include textual and graphical data. Details of these reports are included in Sections 4.6.1–4.6.3.

4.5 Verdict: System architecture and operation

4.5.1 System architecture

Verdict is built around a three-tier architecture model, which comprises the following (see Figure 4.4):

(1) The client tier.
(2) The middle tier.
(3) The database tier.

At the top level of the model is the *clienttier*, which includes the Web browser software (e.g. MS Internet Explorer) that interacts with the Verdict application. In between the top and bottom tiers is the *middle tier*, which communicates data to and from the client and database tiers.

54 e-Business in Construction

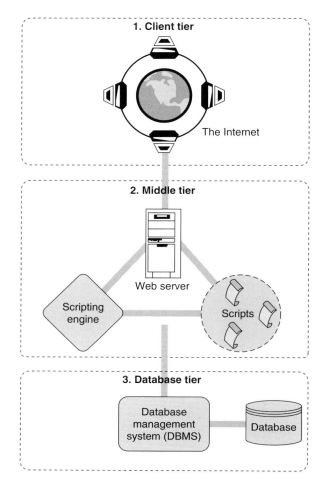

Figure 4.4 **Three-tier architecture of verdict.** (*Source*: Adapted from Williams and Lane, 2002)

The middle tier is more complex and contains most of the application logic. The Web server, the scripting engine and the scripts reside in this tier. The base tier of the Verdict application is the *database tier*, which is made up of a database management system (DBMS) that manages the data that is created, added, modified, deleted and/or requested by the end-user/s.

VERDICT has been developed using PHP (Hypertext Preprocessor) as the scripting language. PHP is an open source[1], server-side, HTML-embedded (Hyper Text Mark-up Language) scripting language used to create dynamic Web pages. It is compatible with many types of databases (Webopedia, 2004). The front-end of the VERDICT prototype is designed using Macromedia Dreamweaver and Fireworks (for graphics).

The application mainly consists of a series of Web-based questionnaire-forms that can be accessed by the end-user/s using standard Web browsers such as MS Internet Explorer and Netscape. Any information that is added to these forms (i.e. end-user responses) is stored in the MySQL database (situated in the *database tier*). MySQL is an open source RDBMS[2] (relational database management system) that can run on UNIX, Windows and Mac operating systems. It has become a popular alternative to proprietary database systems because of its economy, speed and reliability (Webopedia, 2004). The VERDICT system resides on a server with which the end-user communicates. This forms the core of the *middle tier*.

4.5.2 Operation

Any requests made by the end-users are communicated via the Web server. This action invokes the PHP script code embedded in the Web page to request and retrieve the data from the MySQL database. This data, returned by the MySQL database, is then processed by the PHP script. The processed data is then presented to the end-user on a Web page. A high level view of this process, which includes the operations and the system architecture of the VERDICT application are illustrated in Figure 4.5.

End-user companies only require a computer and an Internet connection to access and operate the tool. The performance of the tool will depend on the speed of the Internet connection and the version-type of the operating systems. For example, end-user companies with broadband

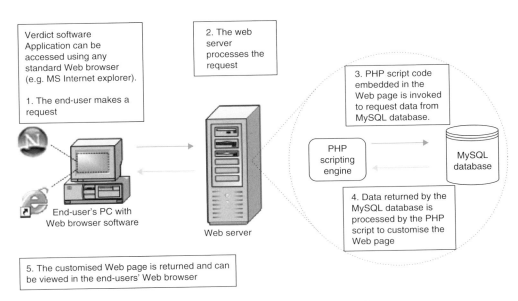

Figure 4.5 Operations of VERDICT. (*Source*: Adapted from Keitz, 2002)

connections will be able to access the site and process information faster than those with dial-up connections. End-users can access VERDICT online using the Web-address: http://civil-unrest.lboro.ac.uk/cvkr.

The next section describes the actual working of the VERDICT software using an end-user case study.

4.6 End-user case study

The VERDICT questionnaire is required to be completed by appropriate company staff for example senior staff with adequate knowledge of the key aspects of the organization – people, process, technology and management. The background information page of this application requires user input in the areas of company information and employee information. The example selected for this case study is that of a large construction management organization with over 500 employees. This end-user company is identified as Company A in this chapter. On completion of the required fields in the background information page, the user begins the e-readiness questionnaire which is distributed over four pages (one page for each category.) A typical page is shown in Figure 4.6.

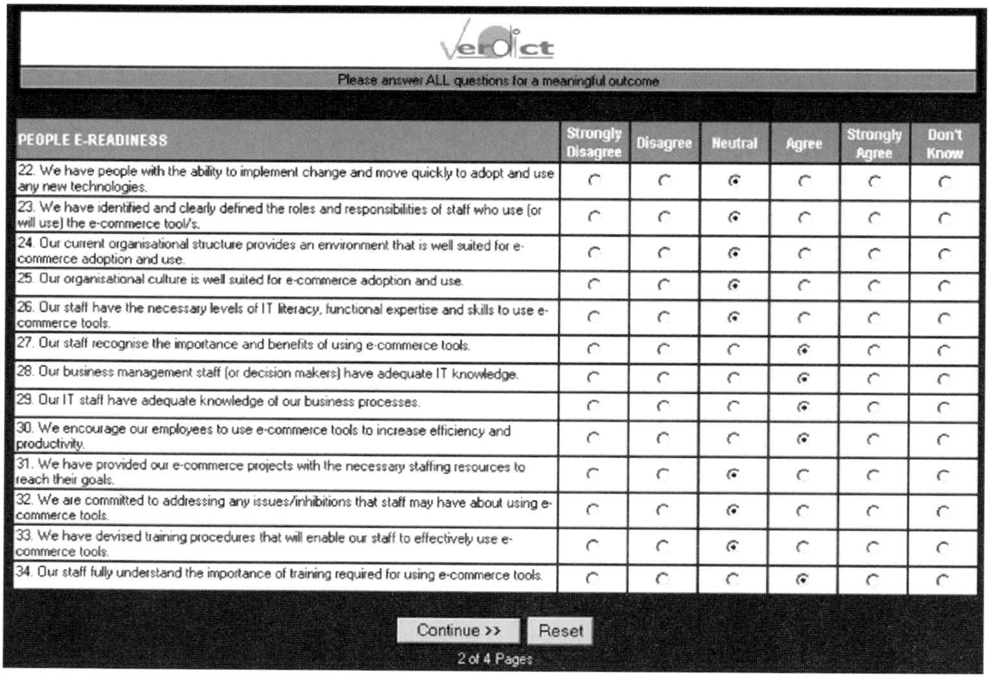

Please answer ALL questions for a meaningful outcome

PEOPLE E-READINESS	Strongly Disagree	Disagree	Neutral	Agree	Strongly Agree	Don't Know
22. We have people with the ability to implement change and move quickly to adopt and use any new technologies.	○	○	●	○	○	○
23. We have identified and clearly defined the roles and responsibilities of staff who use (or will use) the e-commerce tool/s.	○	○	●	○	○	○
24. Our current organisational structure provides an environment that is well suited for e-commerce adoption and use.	○	○	●	○	○	○
25. Our organisational culture is well suited for e-commerce adoption and use.	○	○	●	○	○	○
26. Our staff have the necessary levels of IT literacy, functional expertise and skills to use e-commerce tools.	○	○	●	○	○	○
27. Our staff recognise the importance and benefits of using e-commerce tools.	○	○	○	●	○	○
28. Our business management staff (or decision makers) have adequate IT knowledge.	○	○	○	●	○	○
29. Our IT staff have adequate knowledge of our business processes.	○	○	○	●	○	○
30. We encourage our employees to use e-commerce tools to increase efficiency and productivity.	○	○	○	●	○	○
31. We have provided our e-commerce projects with the necessary staffing resources to reach their goals.	○	○	●	○	○	○
32. We are committed to addressing any issues/inhibitions that staff may have about using e-commerce tools.	○	○	●	○	○	○
33. We have devised training procedures that will enable our staff to effectively use e-commerce tools.	○	○	●	○	○	○
34. Our staff fully understand the importance of training required for using e-commerce tools.	○	○	○	●	○	○

Continue >> Reset

2 of 4 Pages

Figure 4.6 Typical questionnaire page

On successful completion of the questionnaire, users are presented with a report that summarizes their overall e-readiness. This report is divided into the following three sections:

(1) Table summarizing average scores in each category (with 'traffic light' colour coding to indicate e-readiness).
(2) Radar diagram of overall scores in comparison to the 'best-of-breed' in the construction sector.
(3) Summary of all responses highlighting areas that need attention.

4.6.1 Table summarizing average scores in each category

This section summarizes the responses in each category that is Management, People, Process, and Technology and records the average score obtained in each category (see Figure 4.7).

The minimum score that can be obtained for each category equals 'zero' where the respondents 'don't know' the answers to any of the questions, and are therefore not 'e-ready'. The scores are averaged and, depending on the average score, the respondents are presented with 'traffic light' colour coding, that is red, green and amber, to visually indicate their e-readiness in each category, where:

- Red indicates that these aspects need urgent attention.
- Amber highlights those aspects that need attention to ensure e-readiness.
- Green indicates that the end-user organization has adequate capability and maturity in these aspects and therefore is e-ready (in those aspects).

As indicated from Figure 4.7, Company A is e-ready in the areas of people, process and technology. This is represented by the green lights, but the amber light in the area of management indicates that the company has only addressed some of the issues that are required to achieve e-readiness, but still needs to address other issues to ensure e-readiness. This can be

Category name	Average score	Traffic light indicator
Management	3.33	AMBER
People	3.62	GREEN
Process	3.83	GREEN
Technology	4.46	GREEN

Figure 4.7 Typical table summarizing average scores in each category with traffic light indicators

done by focusing on those points in the summary report, where the scores are 3.5 or less.

4.6.2 Radar diagram

Average scores obtained in each category are plotted on a radar diagram. A radar diagram includes 'spokes' which represent dimensions or criteria, scores on which are joined up (Chambers, 2002). This gives the respondents a visual representation of their overall e-readiness (shown with a thick solid line) in comparison to the best-of-breed in construction (shown with a thinner solid line). The radar diagram for Company A is illustrated in Figure 4.8.

4.6.3 Summary report of all responses

A summary of responses to all questions/statements is also included in the final e-readiness report. This includes a list of all the statements included in each category and the corresponding score of each response. This section also highlights specific points within each category that need attention to achieve e-readiness. This allows companies to focus on, and improve on, those specific aspects within each category, even if they may have achieved e-readiness in that category.

4.6.4 Evaluation

Evaluation was based on the functionality of the prototype application, its user-friendliness, and its relevance to its target audience (i.e. construction

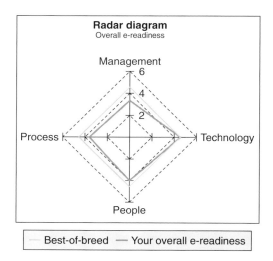

Figure 4.8 Radar diagram illustrating overall e-readiness of respondent's company as compared to best-of-breed

companies). The reviewers (i.e. evaluators) were given a standard evaluation questionnaire covering these areas. Reviewers were encouraged to include any additional suggestions for further enhancing the application.

The VERDICT prototype was evaluated using a number of methods including self-evaluation and peer reviews during the development phase and then through industry validation of the final prototype application.

- *Self-evaluation*: The tool was continually tested for any errors or bugs, which were corrected simultaneously with the development of the prototype.
- *Peer reviews*: A carefully selected panel of researchers and academics conducted the peer review process.
- *Industry reviews*: A random sample of relevant industry practitioners evaluated the prototype application.

The industrial evaluation was conducted by managerial staff of leading UK construction organizations, which included consultants, contractors and project managers. A total of 10 construction companies were approached for evaluating the application. At present five construction companies have stated their willingness to evaluate this prototype. These companies fall in a range of construction disciplines including, contractors, engineers and quantity surveyors. Four of these five have already used the prototype to establish their e-readiness. Table 4.1 shows the average scores of each of the companies in the categories of management, people, processes and technology.

From Table 4.1, it is seen that:

- Management in all the companies is the least e-ready with the lowest scores compared to other three categories.
- All the companies are e-ready in process and technology categories.
- On an average, technology has scored the maximum points (i.e. have a score >3.5) showing that the companies have a high level of e-readiness in this aspect.
- All the companies have 'green-lights' in at least three categories and an amber in the one remaining category.

Table 4.1 Category-wise comparison of e-readiness average scores for end-user companies

Category name	Average scores			
	Management	People	Process	Technology
Company 1	3.48	3.46	3.83	3.54
Company 2	3.48	3.62	3.50	3.85
Company 3	3.33	3.62	3.83	4.46
Company 4	3.52	3.54	3.58	4.69
Average	**3.45**	**3.56**	**3.77**	**4.14**

The initial feedback has been positive and these organizations have welcomed the use of such an assessment tool in construction. One of the reviewers suggested that more needs to be done to help organizations in addressing aspects for which they are not e-ready, however, this is beyond the scope of this project but can certainly be addressed in the future. Other suggestions mainly focused on 'fine-tuning' the application to enhance its user-friendliness and appeal. Such suggestions have been taken on board and the current version of this prototype application reflects this.

4.7 Conclusions and future work

This chapter discussed the outcome of a study that resulted in the development of an e-readiness assessment prototype for the construction sector – VERDICT. The chapter presented the e-readiness model for assessing the readiness of construction organizations for using e-business. The model is based on the premise that for any company to be e-ready, its management, people, process and technology have to be e-ready in order to derive maximum business benefits. The development and operation of the prototype application have also been presented using end-user case studies. The following conclusions can be drawn from the implementation and evaluation process of this tool:

- VERDICT can be used as a self-assessment tool by organizations to gauge their e-readiness (for e-business) in terms of their management, people, process and technology.
- The tool can also help in highlighting areas that need to be addressed to achieve e-readiness.
- Initial evaluation within the industry has shown that VERDICT can be successfully used for evaluating e-readiness of construction organizations.
- Evaluation results show that the case study companies are weakest in management e-readiness and strongest in technology e-readiness (Table 4.1).
- Industry evaluators have welcomed the concept of e-readiness and acknowledged the need for an e-readiness assessment tool, such as VERDICT, to encourage wider adoption of e-business technologies.
- By surveying a large sample of construction sector end-users, the best-of-breed can be established. This can be achieved by following the example of Cisco Systems (2004), who have based their sample on Fortune 1000 executives. A similar approach can be adopted in construction.

The results of this work will be further used to develop appropriate strategies for achieving e-readiness in construction organizations. Besides the positive responses to the evaluation process, some of the reviewers made suggestions that could further enhance the application. Given that this is a prototype application, such improvements will be implemented in future versions of the prototype. Some of the areas that can be improved in the future are:

- Providing guidance to those end-user companies that are not e-ready in order to achieve e-readiness. This can be done by setting short-term achievable targets that lead to e-readiness.
- The relative importance of the criteria for e-readiness of organizations may vary according to the nature of the end-user organization and its discipline. What is a priority for one may not be for another. It will be useful to identify key questions in each category for which the end-users *have to* achieve high scores to be e-ready. Alternatively, provision can be made for the end-users to assign relative weights to the four categories and/or to the individual questions.

These and other improvements can be incorporated in future versions of the prototype.

Acknowledgements

This chapter has been previously published as a paper in *Automation in Construction*, **15**(1), 98–110, January 2006.

Notes

(1) Open source refers to a program in which the source code is available to the general public for use and/or modification from its original design free of charge (Webopedia, 2004).
(2) RDBMS is a type of database management system (DBMS) that stores data in the form of related tables. Relational databases are powerful because they require few assumptions about how data is related or how it will be extracted from the database. As a result, the same database can be viewed in many different ways (Webopedia, 2004).

References

Alshawi, M. and Ingirige, B. (2002) Web-enabled project management. *School of Construction and Project Management*, University of Salford, UK, February, Centre for Construction Innovation, Manchester.

Anumba, C.J and Ruikar, K., (2002) 'Electronic Commerce in Construction – Trends and Prospects', *Automation in Construction*, Elsevier Science B.V. Vol. 11, pp. 265–275.

Basu, S. and Deshpande, P. (2004) Wipro's people processes: A framework for excellence. *White Paper*. Available at: http://www.wipro.com/insights/wipropeopleprocesses.htm (accessed 23rd April 2004).

Bui, T.X., Sebastian, I.M., Jones, W. and Naklada, S. (2002) *E-Commerce Readiness in East Asian APEC Economies – A Precursor to Determine HRD Requirements and Capacity Building*. Available at: http://www.apecsec.org.sg (accessed 17th April 2004).

CERC (1998) *Final Report on Readiness Assessment for Concurrent Engineering (RACE) for DICE* Technical Report, Prepared for the Advanced Research Projects Agency (ARPA) by CERC (Concurrent Engineering Research Center), West Virginia University, USA. Available at: http://www.cerc.wvu.edu/cercdocs/techReports/1993/cerc-tr-rn-93-75/race_body.pdf (accessed 20th April 2004).

Chambers, R. (2002) *Participatory Workshops: A Sourcebook of 21 Sets Of Ideas & Activities*. Earthscan Publications Ltd, London.

Cheng, J., Law, K. and Kumar, B. (2003) Integrating project management applications as Web services. In: Proceedings of the 2nd International Conference on Innovation in Architecture, Engineering and Construction, 25–27 June, Loughborough, UK.

Choucri, N., Maugis, V., Madnick, S., Siegel, M., Gillet, S., O'Donnel, S., Best, M., Jhu, H. and Haghseta, F. (2003) Global e-readiness – for what? *Report of the Group for Globalisation of E-business*, Center for e-Business at Massachusetts Institute of Technology, USA.

Cisco Systems, (2004) *Internet Business Solutions – IQ Expertise, IQ Net Readiness Scorecard*. Available at: http://www.cisco.com/warp/public/779/ibs/netreadiness/20question.html (accessed 15th March 2004).

Construction Industry Times (2002) *Technology: What's in It for Me?* Published by McMillan Scott Plc 2002. Available at: http://www.constructiontimes.co.uk/ (accessed 5th June 2004).

CPA (2000) *E-commerce in the Construction Industry: E-construction*. Construction Products Association, London.

CPA (2001) *E-construction – Where Are We Now?* Second Annual E-construction Survey. Construction Products Association, London.

Craig, T. (2004) *3 Issues to Supply Chain Management Success: Process, People, Technology*. Available at: http://www.webpronews.com/enterprise/crmanderp (accessed 6th May 2004).

Emmett, B. (2002) IT service management: people + process + technology = business value. *The IT Journal*, Third Quarter 2002. Available at: http://www.hp.com/execcomm/itjournal/third_qtr_02/article2a.html (accessed 6th April 2004).

Fuji Xerox (2003) *Aligning People Processes and Technology*. Available at: http://www.fujixerox.com.au (accessed 23rd April 2004).

Goolsby, C. (2001) Integrated people + processes + tools = best-of-breed service delivery. *Getronics White Paper*. Available at: http://itpapers.news.com/ (accessed 23rd April 2004).

Hartman, A., Sifonis, J. and Kador, J. (2000) *Net Ready: Strategies for Success in the E-conomy*. McGraw-Hill, NY, USA, ISBN: 0-07-135242-2.

Hill, D.C. (2004) *Wish I'd Said That!: A Collection of Quotations*. Available at: http://www.wist.info/ (accessed 29th March 2004).

IBM (1999) Arriving at the upside of uptime: How people processes and technology work together to build high availability computing solutions for e-business. *White Paper.* Available at: http://www.dmreview.com/whitepaper/ebizc.pdf (accessed 24th April 2004).

Information Technology and Broadcasting Bureau (2000) *APEC E-commerce Readiness Assessment Guide – A Self-assessment on Hong Kong's Readiness for E-commerce.* Available at: http://www.info.gov.hk/digital21/eng/ecommerce/ec_assessment.html (accessed 20th April 2004).

ITCBP Intelligence (2002) *E-everything for Construction – But What's in It For My Company?* Weekly E-mail Briefing from ITCBP (IT Construction Best Practice), 11th September.

ITCBP Intelligence (2003) *Paperless Office or Still Sifting Documents?* Weekly E-mail Briefing from ITCBP (IT Construction Best Practice), 29th January.

K3 Business Technology Group (2004) *Assessment Tool Checks Readiness for Lean.* Available at: http://www.manufacturingtalk.com (accessed 20th April 2004).

Keitz, F.E. (2002) *Linux Apache MySQL PHP (LAMP) Server Operational Block Diagram.* Available at: http://www.keitz.org/diagrams/kod5001.html (accessed 21st May 2004).

Kern, H., Johnson, R., Galup, S. and Horgan, D. (1998) *Building the New Enterprise: People, Processes and Technology.* Prentice Hall PTR.

Khalfan, M.M.A. (2001) *Benchmarking and Readiness Assessment for Concurrent Engineering in Construction (BEACON).* PhD Thesis, Loughborough University, UK, September.

Kirkman, G.S., Osorio, C.A. and Sachs, J.D. (2002) *The Networked Readiness Index: Measuring the Preparedness of Nations for the Networked World.* Available at: http://www.cid.harvard.edu/cr/pdf/gitrr2002_ch02.pdf (accessed 20th April 2004).

Larkin, B. (2003) *Aligning People Process and Technology.* Available at: http://www.paperlesspay.org/articles/Technology.pdf (accessed 23rd April 2004).

Laudon, K.C. and Laudon, J.P. (2002) *Management Information Systems: Managing the Digital Firm,* 7th edition. Prentice-Hall Inc., NJ, USA.

Maner, W. (1997) *Rapid Application Development Using Iterative Prototyping.* Available at: http://csweb.cs.bgsu.edu/maner/domains/RAD.gif (accessed 7th June 2004).

Ostroff, F. (1999) *The Horizontal Organisation.* Oxford University Press, New York, NY.

Peters, T. (2001) *Comparison of Readiness Assessment Models.* Available at: http://www.bridges.org/ereadiness/report.html (accessed 9th April 2004).

Ruikar, K., Anumba, C.J. and Carrillo, P.M., and Stevenson, G., (2001) 'E-commerce in Construction: Barriers and Enablers', Proceedings of the 8th International Conference on Civil and Structural Engineering Computing, B.H.V. Topping (Editor) Civil-Comp Press, Stirling, United Kingdom, Paper 2 on CD, 2001.

Ruikar, K., Anumba, C.J. and Carrillo, P.M. (2002) Industry perspectives of IT and e-commerce. In: *Proceedings of the 3rd International Conference on Concurrent Engineering in Construction at University of California,* July, Berkeley, 26–40.

Ruikar, K., Anumba, C.J. and Carrillo, P.M. (2004) Impact of e-commerce applications on end-user business processes. In: *Proceedings of the 1st International Conference on World of Construction Project Management (WCPM)* (P.S.H. Poh (chief editor), A.D. Mackenzie and C.J. Katsanis, eds), 27–28 May, Toronto, Canada, pp. 297–311.

Ruikar, K., Anumba, C.J. and Carrillo, P.M. (2005) *End-user Perspectives on the Use of Project Extranets in Construction Organisations*, Accepted for publication in Engineering Construction and Architectural Management, Emerald Press, UK,. Vol. 12, No. 3, pp. 222–235.

Stephenson, P. and Turner, P. (2003) Electronic document management systems in construction: A project-based case study implementation. In: *Proceedings of the 2nd International Conference on Innovation in architecture, Engineering and Construction*, 25–27 June, Loughborough University, UK, pp. 169–179.

Stewart, R.A. and Mohamed, S. (2003) Coping strategies to aid effective information technology implementation in construction. In: *Proceedings of the 2nd International Conference on Innovation in architecture, Engineering and Construction.* 25–27 June, Loughborough University, UK, pp. 69–78.

Sunyit (2004) *Research Methods: "The Big Picture", Telecom Program.* at Sunyit, Tel 598. Available at: http://www.tele.sunyit.edu/rmnote2.htm (accessed 8th April 2004).

The Mosaic Group (1998) *The Global Diffusion of Internet Project*, Mosaic Group. Available at : http://mosaic.unomaha.edu/gdi.html (accessed 21st April 2004).

TUV Management and IT Consultancy (2004) *Management IT Brochure*.

Webopedia (2004) Available at: http://www.webopedia.com (accessed 4th June 2004).

Williams, H.E. and Lane, D. (2002) *Web Database Applications with PHP and MySQL*, Ist edition. O'Reilly and Associates, Inc., CA, USA.

5 Integrated Multi-Disciplinary e-Business Infrastructure Framework

Ihab A. Ismail and Vineet R. Kamat

5.1 Introduction

Early e-business applications in the construction industry were built around focused and narrow processes such as online blue rooms and online project management and collaboration portals. It was estimated that US$2.5 billion has been spent in capital investments, in the United States only, by Application Service Providers creating project management and collaboration portals alone (Nitithamyong and Skibniewski, 2004). Despite the large IT investments completed over the past years, the productivity for many Architecture, Engineering, and Construction (AEC) companies has been flat or has declined in the same period (2005c).

For the construction industry to reap the benefits of such investments, it must equally invest in improving its e-business infrastructure to support the cross-industry deployment and improve its benefit realization rates. Dramatic improvement in industry performance requires equally dramatic changes to business processes and management practices, and, more importantly, a strong infrastructure to support these changes (Severance and Passino, 2002).

e-Business infrastructure extends beyond the physical hardware and routers that support the network connections enabling e-business transactions. e-Business infrastructure is interdisciplinary in nature; it involves business process, trading models, network hardware and software infrastructure, software interoperability standards, legal standards, and more. The ultimate value that the industry can derive from e-business will be reached when those disciplines and standards are aligned in an integrated construction e-business infrastructure. This chapter proposes a framework for the integrated construction e-business infrastructure. It introduces the components of the integrated infrastructure and builds a case for the importance of coordinating the components with each other.

5.2 Integrated construction e-business infrastructure framework

The e-business infrastructure is defined here as the supporting capabilities for online trading between multiple companies which include hardware, software, networks, online payment technologies, security and encryption technologies, online trading business models, legal and regulatory framework, and managerial and organization capabilities. To evaluate the interdisciplinary aspects of construction e-business infrastructure, we propose using a four pillar approach. Figure 5.1 illustrates the skeleton for the proposed integrated construction e-business infrastructure. This chapter will illustrate the importance of coordinating the four dimensions of the proposed integrated infrastructure.

The proposed integrated e-business infrastructure can be broken down into the following four groups of components:

(1) *Technological infrastructure*: Hardware, software, networking protocols, encryption technology, and online payment technology.
(2) *Managerial and Organizational infrastructure*: Business models (Business-to-Business (B2B), Business-to-Consumer (B2C), Consumer-to-Consumer (C2C), Internet-enabled business processes, supply-chain integration capabilities, organizational cultural issues, users training, and return on investment (ROI) justification and calculation.
(3) *Legal and regulatory infrastructure*: Legal requirements for e-business, regulations for electronic trading, and electronic signature act.

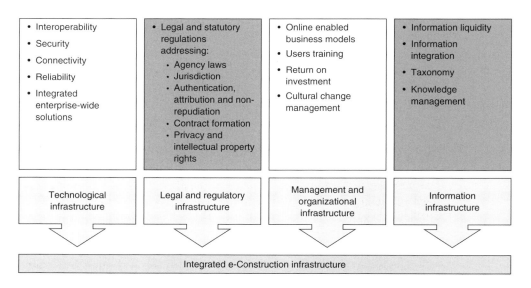

Figure 5.1 Integrated e-business infrastructure

(4) *Information infrastructure*: Includes structures to support information liquidity knowledge management, common across-industry taxonomy, and information availability over the trading network.

Figure 5.2 illustrates an e-business marketplace infrastructure model. It shows buyers and sellers with different connection configurations (single computer access, multiple computer access, and multiple computers, multiple location networked access.) Buyers and sellers can connect to the e-business marketplace through the Internet. Internet connections could be established by connecting to the Internet service provider (ISP) through dial-up connections, digital subscriber lines (DSLs), T1, T4 or other link types. The e-business configuration (e-business marketplace versus single point buyer B2B e-commerce, single point seller B2B or B2C e-commerce) represents the managerial and organizational aspects. The third aspect is represented through the legal and regulatory frameworks within which online trading activities operate, and the fourth aspect represents the market information that flows between different network nodes.

5.3 The importance of e-construction infrastructure

The value of e-business increases exponentially in proportion to the number of users. This is stated in the literature in different variations. For example, Metcalfe's law states that 'the value of a network increases by the square of the number of people or things connected to it', and Kelly's law of networks states that 'in online exchanges, where multiple buyers and multiple sellers come together in a virtual trading space, the potential value of a neutral B2B exchange is thus n^n, where n is the number of users connected to that exchange' (Sculley and Woods, 2001). This is better explained by using e-bay as an example. e-Bay is a Customer-to-Customer (C2C) electronic marketplace which creates many market efficiencies such as easier search tools, better pricing mechanisms derived through auctioning, and an online reputation tracking system; however, the core value of e-bay is its ability to connect more buyers to more sellers than in conventional markets. Gaining critical mass in a market segment is essential to deliver value to participants. Figure 5.3 shows that the growth of gross merchandise sales on e-Bay mimics closely the growth pattern of the number of e-bay users and that both are of exponential nature (source numbers are taken from e-Bay Annual Report) (E-Bay, 2003).

According to Metcalfe's law and Kelly's law, the value of the construction e-business network will increase exponentially in proportion to the number of users. This in turn will drive more members of the AEC industry to join the e-business network putting more pressure on the current e-business infrastructure. Hence, it is important to provide a strong

68 e-Business in Construction

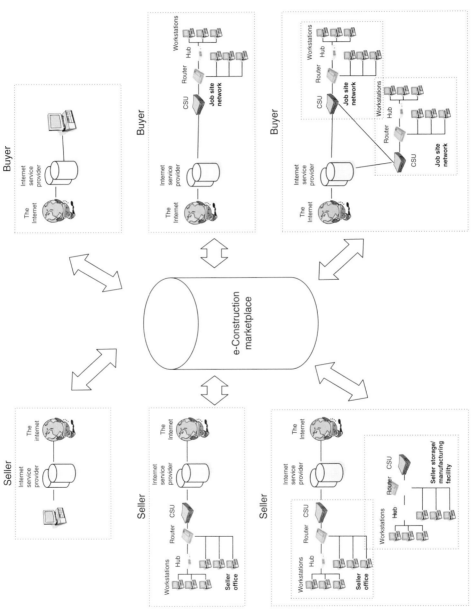

Figure 5.2 e-Business marketplace infrastructure model

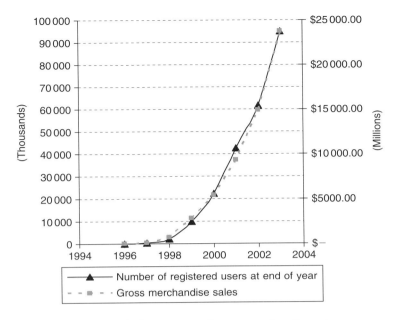

Figure 5.3 Growth of e-bay users and gross merchandise sales

infrastructure to support the growth of e-business and to allow for higher returns on e-business investments.

This growth trend has been identified in the literature for e-business. The time where the construction e-business network moves beyond the early adoption phase was estimated to occur between 2004 and 2005 (Nitithamyong and Skibniewski, 2004). This has been verified by a survey conducted by Engineering News Record (ENR) indicating the number of AEC firms using e-business tools to double every six months (Nitithamyong and Skibniewski, 2004). The discussion of the current challenges facing e-business in the construction industry will be used to illustrate the importance of utilizing an interdisciplinary approach for building a solid e-business infrastructure for the construction industry.

5.4 Summary and status of e-construction challenges

Challenges facing widespread use of e-business in the construction industry is multi-faceted and interdisciplinary. There are challenges under each one of the e-business infrastructure dimensions; additionally, the combination of several issues creates more complex interdisciplinary challenges that are most forgotten. The construction industry in the United States in particular, and in many other developed countries in general, recognized the successes and challenges of e-business. There are many reports in the literature about e-business; many about the challenges of e-business such as the Construction Industry Institute research on e-business (Veeramani *et al.*, 2002), as well as surveys that discuss web-based project management and user view points (Mohamed and Stewart, 2003; Love *et al.*, 2004; Nitithamyong and Skibniewski, 2004; Stewart and Mohamed, 2004). The following section attempts to summarize some of the key individual challenges grouped under each of the four dimensions of the e-business infrastructure framework presented earlier. The discussion is then extended to include additional interdisciplinary challenges that span two or more of the four dimensions of the e-business infrastructure framework.

5.4.1 Technological challenges

Perhaps the one area that is investigated heavily in e-business relates to the technology (Nitithamyong and Skibniewski, 2004). There are several challenges related to the technological aspects of the e-business infrastructure; these are:

- Interoperability
- Security
- Inadequate software
- Connectivity
- Reliability.

Interoperability: The problem of interoperability arises when there is a many-to-many trading relationship. When each supplier is dealing with few buyers, or each buyer is dealing with few suppliers, pre-set exchange protocols can be established alleviating interoperability issues. Interoperability is a major concern in all aspects of e-business. The cost of interoperability is one of the major challenges to wide implementation of e-business (Veeramani *et al.*, 2002). Suppliers are faced with the dilemma of having to adapt to all their client online trading needs while still catering to their own internal standards (Gallaher *et al.*, 2004). Several groups, such as International Alliance for Interoperability (IAI) (2005b) and FIATECH – an industry led, collaborative, not-for-profit R&D Consortium are currently working on solutions for this issue.

Security: Security is also a major concern for e-business applications (Veeramani *et al.*, 2002; Alshawi and Ingirige, 2003; Berning, 2003; Nitithamyong and Skibniewski, 2004). Up until now most e-business applications are using user passwords as the single attribute authentication system. As the industry moves to exchanging more important information online, the single attribute system will no longer be adequate. Also, it will not be adequate for the current use of project management and collaboration and even less adequate for legally enforceable electronic contracts or documents with legal consequences. Current legal requirements, such as the Uniform Electronic Transmission Act requires multi-attribute authentication (Ismail and Kamat, 2006).

Inadequate software: The fragmentation of the software features and the lack of enterprise-wide solutions contribute to the industry's inability to fully utilize the e-business environment to boost productivity (Veeramani *et al.*, 2002; Alshawi and Ingirige, 2003; Berning, 2003). Surveys have indicated that users are looking forward for an enterprise-wide solution instead of the stand-alone software for each business task (e.g. scheduling, estimating, and collaboration) that are currently in use (Veeramani *et al.*, 2002).

Connectivity and reliability: Connectivity and reliability are closely related although different challenges. Connectivity relates to the number of business entities that are connected to a network. As illustrated earlier, the network value increases rapidly with the increase in the number of connected users. The fact that many construction sites are located miles away from the nearest connection point for a T1, DSL, or broad band connection could present a challenge (Fischbach and EC&M, 2002).

Reliability of online e-business applications depends on many factors other than establishing redundancy and reliability at the application server side. To illustrate this further, the network is usually as reliable as its weakest link is. If construction project partners do not utilize an Uninterrupted Power Supplies (UPS) system, they risk connection failures because of power outages even though they subscribe to a network provider that has redundant UPS system in place on their server side. Power failures are only but one cause of network failure. Human error is a common cause of network failures, followed by viruses, hardware failure, software failure, and power failure. Reliability will become more important as e-business goes beyond casual document exchanges to more time-sensitive e-procurement. This will necessitate better connectivity and reliability to avoid unnecessary project delays, especially when submitting time-sensitive data such as online bids and tender documents. Bid recipients on the other side will also want to ensure their systems can handle the high-traffic required at bid submission times (Berning, 2003). Reliability can also surface in different forms other than those previously discussed such as the reliability of the service provide, to remain in business. BuildNet is an example for this challenge. BuildNet filed for bankruptcy and their customers had to switch to other systems when their contracts were bought by BuildNet's competitors as part of the bankruptcy settlement (Berning, 2003).

5.4.2 Legal challenges

The legal challenges are becoming more important and relevant to e-business. The size and complexity of transactions in construction projects presents an important challenge to most e-business activities, which are significantly different from other e-business activities such as buying a book at Amazon.com. The legal issues are agency, jurisdiction, contract formation, validity and errors, authentication, attribution and non-repudiation, and privacy concerns.

The most difficult legal challenge in e-business is posed by intelligent software agents (Ismail and Kamat, 2006). Software agents are computer programs capable of learning and taking decisions on behalf of their owners (Dzeng and Lin, 2004; Ren and Anumba, 2004). They take decisions themselves instead of supporting users in making decisions (Schoop *et al.*, 2003; Ren and Anumba, 2004). Several software agents are developed, or currently in development, that target the construction industry (Anumba *et al.*, 2003; Dzeng and Lin, 2004; Ren and Anumba, 2004; Tah, 2005). Intelligent agents pose serious legal challenges to the current legal system primarily in the area of contracting and agency laws. The basic question is whether intelligent agents could have a separate legal personality from their owner or not? The law will not necessarily accept the capability of agents to take actions on behalf of owners as legal owner consent to bind them contractually (Bain and Subirana, 2003).

Jurisdiction issues pose the next challenge to e-business. Electronic contracts have no geographical boundaries. Laws and regulations, in contrast, do have geographical boundaries. This basic decision of which court has jurisdiction over a dispute is challenged by these contradictions. This challenge often exists in areas like intellectual property rights. Determining jurisdiction for intellectual property rights is important for distribution of drawings and project documents online in e-procurement systems.

A contract is a promise for a promise (Fuller and Eisenberg, 2001). The promises are usually exchanged as an offer and acceptance of the offer. In e-business transactions, however, it may not be easy to distinguish the offeror from the offeree (Ismail and Kamat, 2006). This threatens the essence of the contract for a contract is not formed unless the offer is accepted by the offeree.

Authentication, attribution, and non-repudiation are other set of concerns for e-business. Attributing an electronic message or agreement to an e-contract to the person who purports to send is critical to the success of e-procurement initiatives. The Uniform Electronic Transaction Act (UETA) addresses this issue by allowing electronic signatures *in lieu* of signatures; however, the level of authentication depends on the type of security used to collect the 'electronic signature' (Belgum and Thelen Reid Brown Raysman & Steiner LLP, 1999; Moreau, 1999; Thelen Reid Brown Raysman & Steiner LLP, 1999; Pacini *et al.*, 2002.).

Privacy is also a major concern in e-business. Contractors and owners bidding on e-business marketplaces disclose information about themselves and their bids. e-Commerce marketplaces and portals may derive value from collecting this information (Millard, 2000; Crichard, 2003).

Additionally an owner may be in violation of privacy laws if hackers gain access to confidential bidding information of contractors through their servers.

There is another group of legal risks that are not based on conflicts of law and can be addressed at the contract formation stage for each project. Data ownership after project completion, insurance risks for damages due to electronic system failures, and who bears the responsibility for loss of data or risk of hackers in construction projects using electronic project management systems are few examples in this category.

5.4.3 Managerial and organizational challenges

Managerial and organizational challenges were less emphasized in the literature than the technological challenges at the early stages of e-business implementation in the construction industry. This is contradicting to the fact that almost all industry surveys recognized the importance of those issues (Nitithamyong and Skibniewski, 2004). Recently, however, more studies are focusing on those issues and recognizing their importance:

- Ruikar *et al.* (2006) provided an e-readiness assessment for companies in terms of readiness of their management, people, processes, and technology for the implementation of e-business tools.
- Establishing ROI is difficult for many companies. This in turn creates difficulty in making business case to justify investments in e-business (Nitithamyong and Skibniewski, 2004; Veeramani *et al.*, 2002).
- General strong cultural resistance encountered by all industry sectors. Resistance to change (Veeramani *et al.*, 2002).
- Lack of comprehensive software with features addressing all participants. Current software focus on one entity Contractor, Owner, or Architect (Nitithamyong and Skibniewski, 2004).

Electronic business models used in e-business also pose managerial and organizational challenges. To evaluate why many e-procurement business models pose risks to the construction industry, it is essential to establish the differences between the types of procurement approaches. Procurement contracts can be negotiated or non-negotiated (pre-established). Also, procurement decisions can be based on long-term value maximizing strategic objectives, or short-term single transaction value maximising objectives. Certain e-business models are often more suitable and efficient for one procurement type than the other. Non-negotiated, short-term single transactions are the least complex types of e-business transactions. Often, the one deciding factor for this type of transactions is cost. Negotiated, relation-building transactions are the most complex type of transactions. In this type of transactions, delivery terms, warranty, specifications, and schedules can be all negotiated in addition to price.

The additional complexity warrants additional precautions if the same business model used in less complex transactions is utilized here. Identity and authentication requirements must be stricter and legally binding; pre-qualifications and reputation building protocols also become more important. Even with more precautions and higher system standards, the business model itself could not be suitable to derive enough efficiency to make the model feasible for such transactions. For example, some claim that reverse auctions are not beneficial for long-term market efficiency (Berning, 2003). The business model of reverse auctions work against the market dynamics long-established in the construction industry. The probability that one bidder may bid at a loss in the hopes of making more profit on changes increases the chances of project claims.

5.4.4 *Information liquidity and knowledge management challenges*

To discuss information liquidity and knowledge management and their importance in e-business, a distinction between data, information and knowledge must be first established. Data are 'facts and figures that are meaningful in some way'; Information is 'data that has been organized for a particular purpose' (Blair, 2002). Knowledge, on the other hand, is more difficult to define; however, it could be described as information intertwined with the intellectual ability to put this information to use (Blair, 2002). Knowledge can be either tacit or explicit (Woo *et al.*, 2004). Tacit knowledge is the knowledge experts posses; it is highly context based. Tacit knowledge is further divided into two distinct types: knowledge that can be expressed in logical forms but has not been expressed yet and knowledge that is not expressible (Blair, 2002).

Information liquidity is analogous to financial liquidity. Information has more value when it is more readily accessible, or in other words when it is 'liquid'. Information liquidity is defined as 'the ability to acquire, understand and make use of information when it is needed, where it is needed, in an appropriate context of content and meaning' (Teflian, 1999). Information liquidity is critical to electronic transactions. The value of companies in an all-connected world is shifting more towards its ability to harness the power of information, and growing its experience and knowledge base (Teflian, 1999).

Information liquidity and knowledge management are particularly challenging for e-business. Because of the uniqueness of each construction project, much of the knowledge in the industry is tacit knowledge (Woo *et al.*, 2004). Tacit knowledge is the least liquid type of knowledge and the most difficult to transform into logical statement and information. This poses a great challenge to e-business since the value of the network depends greatly on the amount of accurate information that flows in the network.

5.4.5 Multi-disciplinary challenges

It has been shown that e-business infrastructure is interdisciplinary. e-Business infrastructure is the conglomeration of technological, legal, management and organization, and information infrastructures. One such example of the interdisciplinary nature of e-business is illustrated in the case of Mortenson Co., Inc. versus Timberline Software Corp. is a representative example[1]. Mortenson, a general construction contractor, purchased licensed estimating software from Timberline. The contractor used the software to estimate its construction costs and prepare bids. It specifically used Timberline to prepare for a bid that was accepted and then realized that the bid was $1.95 million less than it should have been.

It was later discovered that there was a bug in the software that Timberline discovered earlier, but thought to be un-important. Mortenson sued Timberline to recover $1.95 million in damages. The court held that the shrink-wrap on the software cover included an agreement that limited the liability of the software company to the purchase price of the software. It expressly waived the right to recover consequential damages. The court also held that the license agreement on the shrink-wrap instructed the purchaser to return the software for a complete refund if they don't agree with the terms. By opening the software pouch, the contractor has accepted a valid way of contracting with a software company and has limited its recovery to the price of the software (Thelen Reid Brown Raysman & Steiner LLP, 2000). However, on the other hand, the court has held the contractor responsible for its submitted price to the owner based on the doctrine of promissory estoppel and on the basis that the owner had relied on the contractor's price.

Promissory estoppel is a long-established legal principal in construction law. For a contract to be enforceable it must have consideration (Fuller and Eisenberg, 2001). For example, a contractor promises not to sell its used equipment that it owns until certain date so that another contractor (potential buyer) arranges its finance, but sell it before that date. The contractor's promise in not enforceable because it lacks consideration. If however, the contractor accepts a down payment, or a certain sum of money in consideration for the promise, the promise is binding and enforceable in court. Promissory estoppel, sometimes referred to as detrimental reliance, is an exception to this rule. Promissory estoppel occurs when 'a promise which the promisor should reasonably expect to induce action or forbearance of a definite and substantial character on the part of the promise and which does induce such action nor forbearance is binding if injustice can be avoided only be enforcement of the promise' (Fuller and Eisenberg, 2001). In American Common Law, Promissory estoppel has been generally utilized in Construction Law to bar bidders from retracting their bids if general contractors depended on their bid price when submitting their bid to the owner. This was justifiable on the basis that contractors had relied on subcontractors' prices, and subcontractors should be responsible for their price quotations given that contractors will rely on quotations for their pricing.

The case illustrates the additional risk borne by the contractor when they used specific software to prepare a bid to submit it electronically. In buying the software, and agreeing via shrink-wrap agreement, that the maximum liability of the software provider is the cost of the software, the contractor indirectly assumed the risk of preparing a faulty bid because of a software miscalculation. The Contractor may not have known that it assumed this risk when it bought the software to prepare the bid electronically.

The significance of this case can be highlighted in the more common scenario where the Owner of a construction project dictates what type of software should the contractor use on any specific project. Owners may elect to do specify certain software to be used to reduce the burden of double data handling and increase the interoperability of the software. The significant point here is that resolving some of the interoperability issues may affect another dimension of the integrated e-business infrastructure, such as the legal risk distribution. This goes to illustrate the importance of having an integrated, interdisciplinary e-business infrastructure that supports the coordination of similar issues.

Another area of coordination occurs between the technological aspects of e-business and the legal aspects. One of the complaints by online project management system users is the large volume of emails received. This problem has been exaggerated by the amount of spam mail that is received in the same inbox. The improvement in technology offered a new solution for this problem in the form of spam filters. User of project management systems can often buy or subscribe to a third party spam filtering system. Such third party spam filters attempt to block the distribution of any spam email to the end-user of project management systems. Sometimes, spam filters mistakenly categorize good emails as spam and block its distribution to end-users. From a legal perspective, it not clear yet what is the liability and risk distribution pattern if a third party spam filter blocks the distribution of an email with some legal significance, such as, for example, a response to a Request for Information from the contractor to architect that proves crucial later in determining liability in a delay claim. Would the contractor be liable for the delay resulting for a miscategorized email that it did not receive instructing it to proceed with some work in the field?

Development of the information infrastructure may also contradict with the development of the technological infrastructure and interoperability requirements. To increase information liquidity some standardization has to happen across the industry. For example, a common taxonomy, and a common product code must be established. On the other hand, interoperability requirements need some standardization in terms of vocabulary and product code structure. Those requirements must be coordinated together or else standardization efforts will fail on both sides. Many other examples can be cited for the importance of an integrated, coordinated e-business infrastructure: security requirements must be coordinated with legal requirements; business models with software requirements; and business models with legal regulations.

5.5 Conclusions

In summary, e-business is poised for widespread application in the construction industry. It is being embraced by more AEC firms everyday. The expected exponential increase in the users is matched with an increase in the amount of investment. Users and the industry will ultimately be looking for e-business to payoff for this investment. This payoff will be derived in terms of added value to the construction operations and more efficiency. To be able to gain this efficiency and value, the infrastructure upon which e-business is built must be strong and well-coordinated. This has not been the case in the past, and is not the case now. The integrated e-business infrastructure framework is ultimately needed to secure this strong foundation and therefore the future of e-business. It is also needed to gain the expected benefits and increase the efficiency of contractors and subcontractors.

Technological challenges have been the primary focus of improvement when it comes to e-business (Nitithamyong and Skibniewski, 2004). This comes as no surprise since it is a topic of highest familiarity for an industry that is accustomed to technical issues and their solutions. The other three groups of challenges have gained little or no attention whosoever. More important, the coordination between the developments of solutions in all four groups does not exist. As the changes are multi-faceted and interdisciplinary in nature, the solution must be interdisciplinary. This gap in the literature is acknowledged and identified in this paper, and is collectively addressed in the current research.

A framework for identifying and categorizing e-business challenges is presented, and an integrated e-business infrastructure framework is discussed. The conflicts between unbalanced improvements in one aspect of the framework and the others are also discussed. Future research is needed to identify different methods to coordinate the work of four aspects of the e-business infrastructure and create linkage between them.

Note

(1) *M.A. Mortenson Company, Inc. (Petitioner), v. Timberline software corporation and Softworks Data Systems, Inc., Respondents.* 140 Wn.2d 568; 998 P.2d 305; 2000 Wash. LEXIS 287; CCH Prod. Liab. Rep. P15,893; 41 U.C.C Rep. Serv. 2d (Callaghan) 357. Supreme Court of Washington.

References

Alshawi, M. and Ingirige, B. (2003) Web-enabled project management: An emerging paradigm in construction. *Automation in Construction*, **12**(4), 349–364.

Anumba, C.J., Ren, Z., Thorpe, A., Ugwu, O.O. and Newnham, L. (2003) Negotiation within a multi-agent system for the collaborative design of light industrial buildings. *Advances in Engineering Software*, **34**(7), 389–401.

Bain, M. and Subirana, B. (2003) Legalising autonomous shopping agent processes. *Computer Law and Security Report*, **19**(5), 375–387.

Belgum, K.D. and Thelen Reid Brown Raysman & Steiner LLP (1999) Legal issues in contracting on the Internet. http://www.constructionweblinks.com/Resources/Industry_Reports__Newsletters/June_1999/june_1999.html (accessed 28 October 2004).

Berning, Paul W. (2003) "E-business and the Construction Industry: User Viewpoints, New Concerns, Legal Updates on Project Web Sites, Online Bidding and Web-Based Purchasing." htp://www.constructionweblinks.com/Resources/Industry_Reports__Newsletters/Dec_22_2003/e_cc(11/01/2004).

Blair, D.C. (2002) Knowledge management: Hype, hope, or help? *Journal of the American Society for Information Science and Technology*, **53**(12), 1019–1028.

Crichard, M. (2003) Privacy and electronic communications. *Computer Law & Security Report*, **19**(4), 299–303.

Dzeng, R. and Lin, Y. (2004) Intelligent agents for supporting construction procurement negotiation. *Expert Systems with Applications*, **27**(1), 107–119.

EC&M (2005a) AEC workers want more flexibility from project management software. Available at http://ecmweb.com/market/electric_aec_workers_flexibility/ (accessed 13 March 2005).

EC&M (2005c) Productivity down, pre-planning up for many construction companies. Available at http://ecmweb.com/mag/electric_productivity_down_preplanning/ (accessed 13 March 2005).

Fischbach, A.M. and EC&M (2002) Team collaborates online to manage airport renovation. http://ecmweb.com/mag/electric_team_collaborates_online/index.html (accessed 1 April 2008).

Fuller, L.L. and Eisenberg, M.A. (2001) *Basic Contract Law*. West Group, St. Paul, Minn.

Gallaher, M.P., O'Connor, A.C., Dettbarn, J.L. and Gilday, L.T. (2004) *Cost Analysis of Inadequate Interoperability in the U.S. Capital Facilities Industry*. Report No. NIST GCR 04-867, National Institute of Standards and Technology, Gaithersburg, Maryland.

IAI (2005b) International Alliance for Interoperability (IAI). Available at http://www.iai-na.org/ (accessed 3 September 2005).

Ismail, I.A. and Kamat, V.R. (2005) Legal risk analysis, modeling, and programming for e-commerce in construction. *Construction Research Congress 2005*, ASCE, Reston, VA.

Ismail, I.A. and Kamat, V.R. (2006) Evaluation of legal risks for e-commerce in construction. *Journal of Professional Issues in Engineering Education and Practice*, **132**(4), American Society of Civil Engineers, Reston, VA, 355–360.

Liebert Corp. (2004) The 2004 Liebert availability report. Available at http://www.liebert.com/support/whitepapers/documents/avail_surv.asp (accessed 3 March 2005).

Love, P.E.D., Irani, Z. and Edwards, D.J. (2004) Industry-centric benchmarking of information technology benefits, costs and risks for small-to-medium sized enterprises in construction. *Automation in Construction*, **13**(4), 507–524.

Millard, C. (2000) *Four Key Challenges for Internet and E-Commerce Lawyers*. Computer Law & Security Report, **16**(2), 75–77.

Mohamed, S. and Stewart, R.A. (2003) An empirical investigation of users' perceptions of web-based communication on a construction project. *Automation in Construction*, **12**(1), 43–53.

Moreau, T. (1999) The emergence of a legal framework for electronic transactions. *Computers and Security*, **18**(5), 423–428.

Nitithamyong, P. and Skibniewski, M.J. (2004) Web-based construction project management systems: how to make them successful? *Automation in Construction*, **13**(4), 491–506.

Pacini, C., Andrews, C. and Hillison, W. (2002) To agree or not to agree: Legal issues in online contracting. *Business Horizon*, **45**(1), 43–52.

Ren, Z. and Anumba, C.J. (2004) Multi-agent systems in construction-state of the art and prospects. *Automation in Construction*, **13**(3), 421–434.

Ruikar, K., Anumba, C.J. and Carrillo, P.M. (2006) VERDICT – An e-readiness assessment application for construction companies. *Automation in Construction*, **15**(1), 98–110.

Schoop, M., Jertila, A. and List, T. (2003) Negoisst: A negotiation support system for electronic business-to-business negotiations in e-commerce. *Data Knowledge Engineering*, **47**(3), 371–401.

Sculley, B.A. and Woods, W.A. (2001) *B2B Exchanges: The Killer Application in the Business-to-Business Internet Revolution*. HarperCollins Publishers, Inc., New York.

Severance, D.G. and Passino, J. (2002) *Making I/T Work*. Jossey-Bass, San Francisco.

Stewart, R.A. and Mohamed, S. (2004) Evaluating Web-based project information management in construction: capturing the long-term value creation process. *Automation in Construction*, **13**(4), 469–479.

Tah, J.H.M. (2005) Towards an agent-based construction supply network modelling and simulation platform. *Automation in Construction*, **14**(3), 353–359.

Teflian, M. (1999). Informaton Liquidity. Cambridge, MA: Time0/Perot Systems. http://www.eyefortransport.com/archive/time03.pdf (accessed 19 September 2004).

Thelen Reid Brown Raysman & Steiner LLP (1999) California is first state in nation to adopt electronic contracting law. Available at http://www.constructionweblinks.com/Resources/Industry_Reports__Newsletters/Nov_18_1999/nov_18_1999.html (accessed 10 May 2004).

Thelen Reid Brown Raysman & Steiner LLP (2000). "Contractor Denied Recovery for $1.95 Million Bidding Error Caused by Allegedly Defective Software". http://www.constructionweblinks.com/Resources/Industry_Reports__Newsletters/Sept_18_2000/defective_software.htm (accessed 20 September 2004).

Thelen Reid Brown Raysman & Steiner LLP (2003) E-commerce and the construction industry: User viewpoints, new concerns, legal updates on project web sites, online bidding and web-based purchasing. http://www.constructionweblinks.com/Resources/Industry_Reports__Newsletters/Dec_22_2003/e_commerce.htm (accessed 1 November 2004).

Veeramani, R., Russel, J.S., Chan, C., Cusick, N., Mahle, M.M. and Roo, B.V. (2002) *State-of-Practice of E-Commerce Application in the Construction Industry*. Report No. 180-11, Construction Industry Institute, Austin, Texas.

Woo, J., Clayton, M.J., Johnson, R.E., Flores, B.E. and Ellis, C. (2004) Dynamic Knowledge Map: Reusing experts' tacit knowledge in the AEC industry. *Automation in Construction*, **13**(2), 203–207.

6 The Role of Extranets in Construction e-Business

Paul Wilkinson

6.1 Introduction

This chapter describes the evolution of construction collaboration technologies – or 'extranets' as they are sometimes called. It draws on case studies and industry surveys and describes how systems are expanding beyond the core area of document/drawing management towards support for various key Architecture, Engineering and Construction (AEC) business processes. For those unfamiliar with construction collaboration technologies, it begins by defining the term. It then looks at the current level of adoption, briefly reviews reasons why the UK AEC industry has been slow to embrace the technologies, and then focuses on the human issues to be addressed in ensuring successful implementation and use of these e-business technologies. It then looks at how the technologies have continued to evolve as vendors work with customers to develop new functionality or take advantage of new technologies.

6.2 Defining construction collaboration technologies

Since the late 1990s, growth of the Internet coupled with the development of better telecommunications links has provided a platform for many new types of information technologies. As this period also coincided with the promotion of more collaborative ways of working within the UK AEC industry – following the Latham report (1994) and Egan report (1998) – there was a demand for applications that engendered collaboration.

In the highly information-dependent AEC sector, many projects are routinely delivered by relatively short-lived, multi-disciplinary, multi-company, multi-location groups of people. IT could be used to send and receive large volumes of project data over longer distances more quickly and cheaply. The proposition was clear: the AEC industry was prone to delay, waste, defects and additional costs due to late, inaccurate, inadequate or inconsistent information. IT offered a way to improve the management of this information; project times, costs and risks could be reduced, and efficiency, communication and quality improved. Thus, from the late 1990s onwards, software developers began to capitalize on the growing AEC demand for more efficient team communication, and the UK AEC sector witnessed the emergence of a group of applications sometimes described, among many other terms, as 'construction collaboration technologies'.

To collaborate, says the Oxford English Dictionary, is to 'work jointly (with) especially at literary or artistic production' (and, perhaps particularly appropriate to the often-adversarial atmosphere of traditional construction projects, it adds: 'to cooperate with the enemy'). However, the term 'collaboration technology' may, arguably, be something of a misnomer. If collaboration is essentially regarded as a creative process or capability reflecting the roles and responsibilities of the participants concerned, it is clear that it is people that collaborate, not technologies or systems; as the Butler Group (2003) succinctly put it: 'Collaboration is an activity – not a piece of technology'. Using so-called collaboration technologies does not necessarily mean one is 'collaborating'; it is, in short, an enabler, a platform that allows collaboration to take place when people are prepared and equipped to do so.

Individual collaborators need a point of access where they can participate in a process, sharing their collective information with other team members. Collaboration involves an interface – a shared environment where processes and information can be efficiently and effectively integrated. Numerous different synonyms and abbreviations have emerged to describe this shared environment (Wilkinson (2005) lists more than 20 different terms), some inappropriate or potentially confusing. The term 'project extranet', however, became one of the most widely used. Fitting somewhere between private internal Websites ('intranets') and conventional public Websites, 'project extranet' did at least convey the key notion that information could be shared privately between project team members drawn from several separate organizations. In 2003, the founder members of the UK vendors' trade association eschewed the word 'extranet', opting instead to become the Network for Construction Collaboration Technology Providers (NCCTP).

Broadly, all such systems can be accessed through a computer equipped with a standard computer browser (e.g. Microsoft Internet Explorer) and a working Internet connection. They share the same basic functions. Authorized users, irrespective of their geographic location, can get immediate access to a single shared central repository of project data that grows as information about the project or programme (a building, a bridge, a water treatment plant, etc.) is developed by the team. Feasibility studies, budgets, sketches, drawings, approvals, schedules, minutes, photographs, specifications, standards, procedures, virtual reality models and other project-related documents can all be viewed; team members can add comments, issue notices, instructions and requests for information (RFIs), and publish drawings and documents, either singly or in batches. The users of these systems can work on the most up-to-date, accurate and relevant information, backed by all the archive material.

On a typical construction project of medium scale (with a capital value in excess of £5 million or US$10 million) or above, there will usually be dozens of participants, hundreds of drawings and documents, and thousands of information exchange processes. The core project or programme will often be sub-divided into many smaller packages, phases and/or contracts; similarly, some team members may only be involved in the completion of relatively small tasks; as a result, these individuals may not share the main goal of the project, but their role-specific goals should at least contribute towards the achievement of that main goal. Kalay (1999) pointed out how construction collaboration differs from that in other fields:

> First, it involves individuals representing often fundamentally different professions who hold different goals, objectives, and even beliefs. Unlike collaborators in medicine or jurisprudence, who share a common educational basis, architects, structural engineers, electrical engineers, client, property managers, and others who comprise a design team, rarely share a common educational foundation, and often have very different views of what is important and what is not. Second it involves ... 'temporary multi-organisations': a team of independent organisations who join forces to accomplish a specific, relatively short-term project. While they work together to achieve the common goals of the project, each organisation also has its own, long-term goals, which might be in conflict with some of the goals of the particular project, thereby introducing issues that are extraneous to the domain of collaboration. ... Third, collaboration in the building industry tends to stretch out over a prolonged time, even when the original participants are no longer involved, but their decisions and actions still impact the project. (p. 4)

Our definition, therefore, needs to cover some of the principal features of the technology and to allow for these multiple goals.

'Construction collaboration technology' is:

> A combination of technologies that together create a single shared interface between multiple interested individuals (people), enabling them to participate in creative processes in which they can share their collective skills, expertise, understanding and knowledge (information), and thereby jointly deliver the best solution that meets their common goal(s), while simultaneously creating an auditable electronic record of the people, processes and information employed in the delivery of the solution(s).

6.3 Uptake of construction collaboration technologies

By the mid-2000s, construction collaboration technologies were only slowly becoming a normal part of the project delivery process in UK construction projects. For many project teams, even email – almost ubiquitous in other sectors – took a while to become accepted; Autodesk's 2003 UK survey found that, while 86% of its respondents said they used IT to communicate regularly or most of the time, for over 60% this meant the regular use of CD-ROM and post – not email – to distribute data to third parties. In 2004, Davies conducted the first of two surveys of CAD managers, finding project extranets accounted for 2.9% of data issues. Two years later (Davies, 2004), he found that 'Email continues to be the most popular form of issuing digital information, at 65.5%, [but] project extranets are now the primary issue format of 13.8% of companies.' In the space of two years, then, according to Davies (2004, 2006), the proportion of digital information published to construction collaboration systems grew four-fold, but there was clearly still some way to go if they are to seriously challenge the industry's reliance on other communication routes.

The picture painted by other surveys over the past few years is variable. A *New Civil Engineer* poll (Hansford, 2002) said 38% of respondents were using online project collaboration. Barbour (2003, p. 14) said 13% of the 322 client respondents to its telephone survey claimed their teams used such technology, rising to one-third of those spending over £100 million. In June 2004, the IT Construction Forum surveyed 373 firms of all types and sizes, and 43% of its respondents said they used project extranets to collaborate online. The DTI benchmarking study (2004, p. 52) found 17% of construction businesses claiming to be extranet users.

While the proportion using the technologies is growing, why has there been only partial adoption? Respondents to market research undertaken by the NCCTP in 2006 identified several barriers or disadvantages that were hindering adoption (see Figure 6.1).

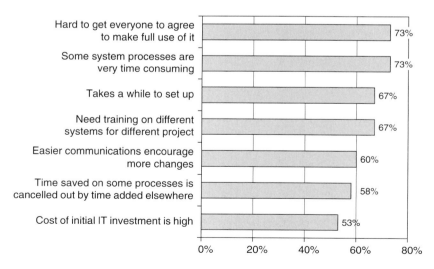

Figure 6.1 Main barriers/disadvantages encountered with extranets. (*Source*: From NCCTP (2006), *Proving Collaboration Pays*)

Additional issues include:

- Small projects where use of collaboration technologies would be inappropriate – while the technologies have clear appeal for any project of a significant scale, complexity and budget, many construction projects are small and relatively simple affairs requiring only a few drawings and other documents to be shared among a few individuals.
- The relative newness of the technologies – adopted first by 'innovator' clients, contractors and consultants within a largely conservative industry, less progressive individuals and organizations remain wary.
- The immaturity of the market – the technologies suffered by association with the boom and bust of the dot.com era. Buyers were wary of vendors with short track records and/or uncertain futures, of systems or architectures that might not prove to be 'industry standard', and of Internet-based solutions over which – as they saw it – they had less control.
- Some 'techno-phobia' – AEC organizations can be wary of taking on the role of information manager for a project team, and may have 'hosting hesitation': concerns about using an externally hosted solution. Some teams have found it difficult to choose between different systems, have been wary about confidentiality, or worried about replicating common project processes. There have been 'connectivity concerns' (do I need to update my existing software, hardware and telecommunications systems?), and concerns about using electronic as opposed to paper-based communications, including data ownership, copyright and access to data in case of disputes.
- The need for multilateral commitment – unlike selecting a technology for use within a business (word-processing, CAD, cost control), the adoption and use of a collaboration system tends to require pan-disciplinary commitment and involvement and, often, the active involvement of the client.
- Lack of clear evidence of the business benefits of the technologies – another factor is the desire of many potential buyers to get concrete evidence that the technologies will deliver time savings, cost savings, efficiency improvements, improved competitiveness, etc.

This final point was perhaps the most important issue, and also the most difficult to crack. To date, most of the evidence offered by technology vendors has tended to be anecdotal. Few vendors (or their customers or project teams) had consciously set out to measure the benefits – indeed, the DTI (2004, p. 95) found construction businesses were the least likely of all sectors to measure the costs and benefits of technology. Influential American analyst Joel Orr (2002, pp. 11–12) believed there were several reasons behind this lack of research into overall productivity improvements:

- To speak of 'productivity increases' requires careful measurement of productivity prior to implementation of the new technology. This is

difficult to do in the project-oriented world of construction, and companies do not seem motivated to do it.
- Some issues are self-evident, and construction professionals do not want to invest time in proving them – for example, the fact that electronic transmission of documents is much faster and cheaper, and more auditable, than using courier services.
- For many of the parties to a construction project, productivity is not a clearly defined concept. To put it bluntly, if one is being paid by the hour, reducing the number of hours required to get a job done is not an attractive proposition…. Only the owner is clearly motivated to do more with less….
- Most … vendors underestimate the extent of computer illiteracy in the construction community, and thus underestimate the amount of training required for successful project implementation.
- Construction projects are not highly disciplined affairs. Unless the use of a new tool can be tied to payment, subcontractors will tend to do things 'the old familiar way', despite any benefits they might gain from the new tool.

Similarly, some industry professionals often insist that every construction project is unique, and that the benefits gained from one project will not necessarily be experienced on the next. In the same vein, different benefits might be achieved through use of different collaboration systems; some benefits might apply across all the main systems – others might only be delivered by particular individual applications; and some benefits might only be achieved because the project team also had an impact. For example, Howard (2002) describes a project undertaken by UK airports operator BAA plc in which overall 3% cost savings were achieved due to use of collaboration software and partnering framework deals – but, overall, 70% of the project savings were attributed to the technology.

Finally and perhaps more importantly, industry businesses may be reluctant to admit that they can sometimes be inefficient, that activities can be duplicated, that tasks need to be redone, etc. Drawing on past experience, many project budgets and schedules are prepared with healthy contingencies to allow for unplanned events, some of which often arise simply because of communication breakdowns between project team members. Detailed measurement of the use of collaboration technology on a project might expose such inefficiencies.

6.4 Benefits of construction collaboration technologies

However, there has also been some research into the benefits. While Wilkinson (2005) identified and discussed a handful of studies, most were limited in scope (small samples, limited investigation) and their findings correspondingly inconclusive. However, a landmark study was published in early 2006 by researchers at Harvard University. Becerik and Pollalis (2006) used detailed case studies and questionnaires to assess the benefits of what they termed OCPM (online collaboration and project management) technology. They focused on three categories: tangible, quasi-tangible and intangible benefits.

6.4.1 Tangible benefits

Tangible or quantifiable benefits such as cost reductions and time savings are usually measurable in monetary terms. Vendors will often claim that employing their technologies will reduce expenses and speed up processes. For example:

- Reductions in expenses, materials and man-hours relating to printing, reproduction, distribution, storage/archiving, management and retrieval of drawings, documents, photographs, forms, etc.
- Reductions in travel, meeting and fax/telephone costs.
- Less time spent searching for or chasing already-existing information, or working on out-of-date information.

In their research (2006), Becerik and Pollalis focused on tangible cost savings associated with e-RFIs, e-bidding (i.e. online tendering) and electronic document transfer. On RFIs, they found the use of OCPM technology reduced RFI turnaround time, reduced the time spent on dealing with RFIs (they calculated an annual saving in staff time costs of $13 457 per project) and produced savings in paperwork and transfer equivalent to $694 per project. On bidding, they calculated that one organization could save $52 520 a year in staff and paperwork-related costs. On savings arising from using electronic document transfer, Becerik and Pollalis calculated an annual saving of $53 664 due to reduced reliance on Fed-Ex by a construction client business; even making quite modest assumptions about printing and copying savings, pointed to a potential saving of $4875 annually per project (pp. 26–32).

The NCCTP research study (2006) asked respondents to compare how long it took to approve drawings exchanged via a collaboration solution as against those exchanged by email or other means: survey respondents said average drawing approval times were reduced by 26%, from average of 9.3 days to 6.9 days.

6.4.2 Quasi-tangible benefits

Quasi-tangle benefits tend to relate to efficiency improvements that are quantifiable, but difficult to measure. Again, there is ample evidence embedded in vendor case studies, but relatively little independent verification of these benefits. However, Becerik and Pollalis (2006) researched 27 potential quasi-tangible benefits, asking respondents to rank the benefits on a scale from 1 ('very low') to 5 ('very high'); see Table 6.1.

Table 6.1 Quasi-tangible benefits

Average ranking (out of 5)	Quasi-tangible benefits
4.35	Improved data availability
4.19	Enabled having complete audit trail
4.00	Improved information management
4.00	Enabled faster reporting and feedback
3.97	Provided accurate and timely information to give valid/accurate decisions
3.95	Improved process automation (RFIs/change orders, auto-updated master budget)
3.93	Improved information version control
3.84	Enabled better project/program control
3.61	Improved timely capture of design/construction decisions
3.57	Enabled fewer information bottlenecks
3.56	Enhanced working within virtual teams
3.47	Enabled quicker response to project status and budget
3.41	Improved quality of the output
3.29	Enabled better forecasting and control
3.26	Improved project relationships with strategic partners
3.20	Reduced rework/data re-entry
3.06	Enabled better resource allocation; more effective assembly of project teams
3.05	Improved public relations
3.03	Reduced personnel costs due to improved efficiency
2.94	Improved idea sharing among team members/within organization
2.94	Minimized project/business risks
2.91	Enabled faster launch to market due to faster delivery
2.88	Reduced errors or omissions
2.87	Reduced delivery lead teams
2.75	Enabled better inventory management
2.56	Enabled more effective identification and assessment of new suppliers
2.38	Enabled advance purchase of materials

In performing this research, and particularly by proposing this methodology, Becerik and Pollalis have performed a major service for the construction collaboration technology sector. The methodology could be applied by other researchers, and by vendors on internal surveys of their users – the NCCTP survey of users of its members' collaboration systems (NCCTP, 2006) incorporated some of its qualitative approach, but also presented respondents' opinions in a more quantitative manner (see Figures 6.2–6.4).

The Role of Extranets in Construction e-Business 89

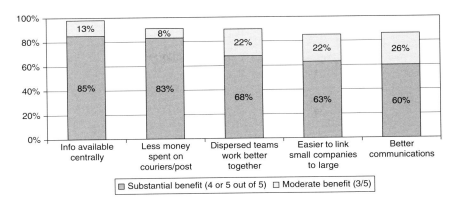

Figure 6.2 Main benefits experienced in project management, communications and team working. (*Source*: From NCCTP (2006), *Proving Collaboration Pays*)

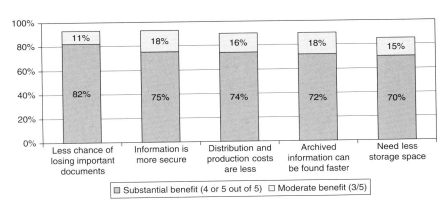

Figure 6.3 Main benefits experienced in document management, storage and retrieval. (*Source*: From NCCTP (2006), *Proving Collaboration Pays*)

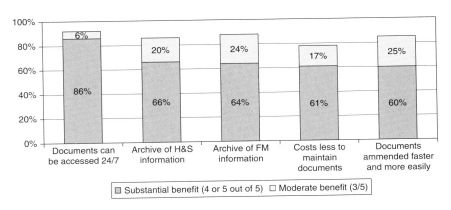

Figure 6.4 Main benefits experienced in hand-over, commissioning, operations and maintenance. (*Source*: From NCCTP (2006), *Proving Collaboration Pays*)

6.4.3 Intangible benefits

Becerik and Pollalis (2006) define intangible benefits as the level of new outputs enabled, and were discussed qualitatively. From their case studies, they identified several qualitative benefits including improved internal knowledge management, process and workflow reengineering, supply chain integration, development of new business, improved forecasting, claims mitigation and management, and improved performance measurement. The NCCTP (2006) found evidence that construction collaboration technologies also helped differentiate contractors with relevant experience from those without (see Figures 6.5 and 6.6).

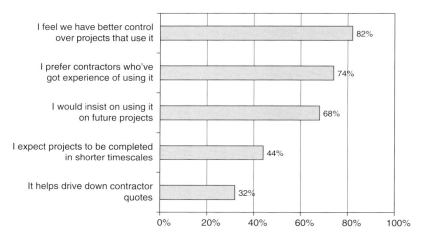

Figure 6.5 Client opinions relating to collaboration technologies. (*Source*: From NCCTP (2006), *Proving Collaboration Pays*)

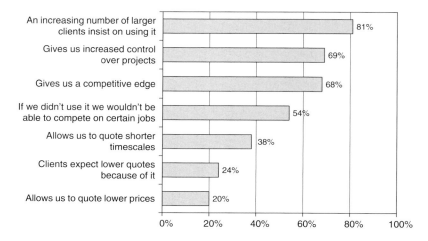

Figure 6.6 Non-client opinions relation to collaboration technologies. (*Source*: From NCCTP (2006), *Proving Collaboration Pays*)

6.5 Human aspects of collaboration

Experience in other industries (e.g. IT, retail, manufacturing) suggests that failure to understand and adapt human behaviour, rather than technology, is the biggest single impediment to successful collaborative working. Arguably, in the more slow-changing – even change-resistant – AEC industry, culture or people issues are even more important:

> ... not a single project Website helps establishing ground rules, building team spirit, trust and commitment, takes in consideration the experience or seniority of individual team members, or accounts for the external social and business environment behind each team member (Verheij and Augenbroe, 2001).

As mentioned, adoption and use of a collaboration system tends to require pan-disciplinary commitment and involvement and, in many instances, the active leadership and encouragement of the client. In short, it is not usually a technology issue. Indeed, to echo points made repeatedly by vendors, commentators, conference speakers and users since 2000:

(1) Collaboration requires a combination of people, processes and technology/information.
(2) Successful collaboration is 80% people and processes and 20% technology/information (some say it is more like 90/10 – but the stress is always on the people and process issues).

In other words, successful collaboration is much more dependent on the culture of the team than it is on the technology it employs. While the various vendors may like to describe their systems as collaboration technologies, they will not enable collaboration at all if the project team is unable or is unprepared to collaborate. There may still be benefits in terms of faster communication, easier access to a central repository, reduced paperwork costs, etc., but more substantial savings of the kind envisaged by Latham and Egan, and promised by the technology providers, will only come from rethinking and changing the culture of the team and the processes it employs.

A key factor in this will be leadership. Emmett (2002 – cited by Ruikar et al., 2006) stresses that 'the people, processes, and technology need a leader', just as 'an orchestra needs a conductor':

> In an orchestra.... You've got musicians (people); musical scores (processes) and musical instruments (technology). But without a conductor, they're more likely to produce noise than music. Even if everyone in the string section plays the right notes at a relatively similar tempo, creating a symphony requires more than following the sheet music.

In other words, as Ruikar *et al.* (2006) argue: 'for an organisation to be e-ready it must have … management that believes in the technology and takes strategic measures to drive its adoption, implementation and usage in order to derive business benefits from the technology' (p. 101).

The resistance to the technology can be analysed by reference to two broad areas: resistance to the strategic principle of collaborative working, and tactical resistance to the adoption of the technology itself (NB: apparent concerns about the technology can often be used to disguise deeper issues about collaborative working as a whole).

6.5.1 Resistance to collaborative working

Despite Latham and Egan exhorting the AEC industry to adopt more collaborative, integrated working methods, industry estimates suggest that the majority of UK projects are still procured using more traditional, often adversarial approaches. This partly reflects inertia in contemplating more open, collaborative approaches, whether this is found at an individual level, departmental or intra-organizational level, inter-organizational level or within the industry as a whole. Each of these dimensions will be briefly examined.

Individual resistance

In AEC firms, individual advancement has frequently depended upon gaining professional qualifications and years of project experience, using familiar, traditional, tried and trusted techniques. It can be very difficult to persuade individuals that they need to change, to adopt a different mindset and behave differently (and to use a new technology into the bargain), particularly if their entire careers have been devoted to achieving seniority through age and continual demonstration of their individual expertise and experience.

Attitudes to collaboration will also vary from profession to profession. Some workers will be well attuned to working in teams; others will see themselves as individuals. Verheij and Augenbroe (2001), for example, contrast the different approaches of contractors and designers, characterizing the former as 'process workers' and the latter as 'knowledge workers'. For process workers, the team concept is simply part of their regular work environment, while in knowledge work, the concept of team is frequently associated with a loss of creative freedom and individuality.

Moreover, the sometimes confrontational or adversarial nature of traditional UK projects can preclude genuine teamwork, as participants seek to avoid risk or blame should things go wrong. Older designers who may always have worked in such a way might seek to justify why they should not change (see Box 6.1), and will need active encouragement and support to make the necessary changes.

> **Box 6.1 Online mark-up or 'red-lining'**
>
> A common feature within most construction collaboration platforms enables users to make comments on a design drawing online. Typically, a user might draw a circle or a cloud around an element of the design and ask a question or give an opinion about it. This mark-up or 'red-line' is then saved and permanently associated with the relevant drawing, communicated to other people and becomes part of the drawing's audit trail.
>
> In BIW's and other vendors' experience, this feature is not always universally welcomed. For example, BIW and 4Projects formed a NCCTP delegation which met with the Major Architects IT Group of the Construction Industry Computing Association in November 2005. The topic of red-lining came up: 'Red-lining is a pain.' ... 'We don't just make a few comments, we bleed over drawings!' ... 'It's almost impossible with a mouse, and digital pen devices and tablets are an expensive extra.' ... 'We would need larger screens but they are so expensive ... and the screen resolution isn't good enough ...' One firm described how teams print out their drawings, mark them up by hand and then 'get a monkey to enter them onto the extranet system'.
>
> At the same time, there is an opportunity for those teaching younger or new industry professionals (educational institutions, professional bodies, mentors, etc.) to begin to promote more collaborative attitudes and approaches. As Latham (2004) sums up: 'It needs serious training, deep culture change led from the top and continuous reinforcement.'

Intra-organizational resistance

Individual progress towards more collaborative approaches to their activities will count for little if their employers do not also encourage and support such approaches. Within many organizations, there can be departmental resistance to the notion that they should share information. Key functions – sales, IT, procurement, HR, accounts, etc. – often sit in 'silos', with their own agendas, systems, attitudes (including the often destructive 'not invented here' syndrome) and varying degrees of influence over corporate strategy. There may also be regional silos within which different parts of an organization pursue regional agendas that differ from each other, and from that of head office. However, if a business can begin to identify where, why and how collaboration might deliver business benefits, it can begin to justify making some quite profound changes.

Organizations need to understand and develop the collective experience, knowledge and wisdom of their own people, and then to devise and implement supporting cultures, structures, systems and technologies. For example, managers could amend employee job descriptions to emphasize team performance and, while accepting there is still room for individual brilliance, place less emphasis on individual achievement alone. Senior managers ought to be seen to preach and practice collaborative working. Collaborative working should be rewarded, thus motivating and incentivising employees to change their attitudes and behaviours. The objective, according to Davis (2003, p. 22) is to: 'to create an environment where people are not only comfortable, but also positively enthusiastic about collaboration'.

In short, to avoid accusations that they are only paying lip service to the principles of collaborative working, organizations must resolve any internal issues they have about collaboration before they start considering committed, collaborative relationships or alliances with external clients, partners or suppliers. Once these intra-organization issues are eradicated, focus can switch to breaking down the inter-organizational fear and mistrust that often exists.

Inter-organizational resistance

Traditionally, project participants in the UK AEC industry have established external trading relationships based on short-term commercial outcomes relating only to the immediate project. Essentially, the approach was adversarial, focused on cutting costs/maximizing profits from the transaction, while minimizing defects and delivering the project on time, with onerous contracts setting out participants' roles and responsibilities and managing their risks.

But collaborative working has begun to drive change. Hierarchies are eroded and, instead of the traditional handover of responsibilities (e.g. from architect to contractor to facilities manager), all parties increasingly take an interest in the whole process of delivering the project. There is also increasing emphasis on longer-term relationships, helping partners to focus on working together and delivering mutually acceptable levels of profit over a series of projects, and on achieving and sustaining improvements in design, service quality, health and safety performance and innovation. The desired benefits also include more qualitative 'relationship' factors such as improved cooperation, fewer disputes or claims and better communication (see Box 6.2).

> **Box 6.2 Andover North Redevelopment (IT Construction Best Practice Programme, 2003)**
>
> The UK Ministry of Defence took a lead in adopting partnering-type approaches – such as 'prime contracting' – for its projects, and one of its first such projects was the Andover North Site Redevelopment, where Bucknall Prime Solutions (formerly Citex Prime Solutions) was awarded the contract.

> From the outset, great emphasis was placed on working closely together. To support the transparent partnering culture, the team agreed to use a Web-based collaboration system from BIW to provide a communication platform to share and exchange all documents, drawings and other information relating to the project. This agreement was written into a key document, the project execution plan, endorsed by the client.
>
> The extranet was populated with the latest revisions of all drawings and relevant documents and key users attended a two-day course that equipped them to train other users within their companies. As training extended across the project team, usage grew rapidly, averaging more than 2000 log-ins every month throughout the peak period. In 20 months, there were over 23 600 system log-ins. 1733 documents and 4072 drawings were published, over 1300 comments on these were made, and all were accessible to over 170 users from 27 organizations.
>
> Instead of designers distributing multiple packages of drawings, drawings were published once – to the BIW platform – and individuals could then view, comment upon and, if necessary, print off just the drawings or details they needed. This cut the volume of paperwork produced, distributed and stored. Team feedback suggests this reduction was also partly due to the partnering ethos as the absence of an adversarial culture removed the need for the many contract letters found on more traditional projects. Information could be found more readily. As well as managing information processes during design and construction, the extranet also created an efficiently searchable electronic archive of all information relating to the buildings, for future operation and maintenance purposes.

While collaboration may require organizations to adapt their internal structures, cultures and managerial processes, the change may also extend to how they encourage and support – instead of block or stifle – collaborative processes across organization boundaries. For example, organizations might consider how employees could be encouraged to work as staff of a 'virtual company', perhaps incentivized on the extent to which they contribute to the mutual objectives and success of that entity (the Bucknall Austin-led prime contracting team involved at Andover North (see Box 6.2) subsequently created a new organization, Novus Solutions Ltd, to target other customer opportunities demanding similar skills and experience).

Industry resistance

Many of the issues raised in the previous three sections contribute to an overall industry resistance to the notion of collaborative working. Much of the UK AEC industry is 'heavily shackled by a conservative culture and mindset [and] inhibited, in many cases, by organisational inertia' (Autodesk, 2003, p. 1); it is very fragmented, with the danger that collaborative working becomes the norm for a progressive minority, while the majority of the sector continues in its old ways; and most project teams tend to be focused on design and construction, not on the 'whole life' of a built asset. While there have been some clear examples of the benefits of partnering, the approach has yet to become the accepted norm across the UK construction industry.

6.5.2 Resistance to collaboration technologies

This section considers the UK AEC industry's attitude to technology, and to collaboration technologies in particular.

The industry has been keen to use new IT tools where appropriate, and is not completely techno-phobic. As with most industry sectors, attitudes to information technology vary. Autodesk's 2003 industry survey suggested there was a division on future strategy between 'long-termists' who see quality, collaboration and effective partnerships as primary drivers and 'short-termists' who focus on cost and immediate return, with the latter more common. While most respondents recognized the need for change (e.g. partnering and integration), they were unsure about implementation, doubtful about who stood to benefit most, and – apart from IT – negative about all the supposed facilitating factors, especially culture and mindset: 'the industry's heritage and innate conservatism is, they confirm, a source of major weakness' (p. 5). Looking more specifically at extranets, Croser (2003, p. 73) suggested there was 'fear, uncertainty and doubt (FUD) somewhere between apathy and hesitation'.

These cultural barriers to technology adoption may also relate to the age of many decision-makers, as Becerik (2004, p. 241) identified. Attitudes to technology can also vary according to the size of the AEC organizations involved. Looking at contractors, for example, Stratagem/DTI (2003) found project collaboration was an important e-business issue for 57% of companies with more than 100 employees but was important to only 35% of companies with between 50 and 100 employees. Their study highlighted: 'a large gap between large main contractors activities in e-business, (91% use email, 58% project collaboration), with their supply chain and small subcontractors ability to respond, (only 48% use email and 12% project collaboration)' (p. 28).

Naturally, one might also expect attitudes to technology to change over time. As particular technologies become more accepted and more commonplace – as they make the transition from 'novel' to 'normal' – their take-up by construction industry organizations will also increase. As far as adoption and use of Web-based construction collaboration technology is concerned, this transition will require individuals and organizations to resolve some of the technology issues already mentioned ('dot.com doubt', 'hosting hesitation', etc.).

6.5.3 Managing the human/technology issues

Several practical organizational issues still remain if a project or programme team is to successfully select and implement construction collaboration technologies, including:

Selection

Prolific client organizations are beginning to adopt single construction collaboration solutions that can be employed across all their schemes, but many consultants and contractors, while sometimes favouring particular systems, will still tend to follow any stipulations made by a client, even if it means staff need to be proficient in the use of several systems at once. To help encourage later buy-in and take-up of the selected system, it is common to involve as many members of the project team in the selection process as possible. Industry references may be taken up, and a 'due diligence' process may also be undertaken to check, say, a vendor's hosting facilities, its financial stability and the legal protections built into contracts and service level agreements, etc.

Timing

Industry experience to date suggests that the best time to introduce collaboration technology to a construction project team is as early as possible – usually the point when team members are appointed and fees are agreed to move the scheme forward from the conceptual stage towards a more detailed consideration of design and buildability issues, that is when the initial project team begins to expand significantly and include individuals from other disciplines and/or companies with whom the core team need to collaborate. Team members can then get involved in the 'nuts and bolts' of how the technology will be configured to suit the project and their inputs to it.

Encouraging buy-in

Involving end-users in the selection process is a key stage in achieving buy-in to the adoption and use of a system. The switch from conventional paper-based information systems to ones employing shared centralized repositories of project data can seem a major step. This may be a 'generational' or 'Luddite' attitude, with older and less IT-literate professionals resisting new methods, while their younger, more flexible and more IT-literate counterparts embrace it (see Box 6.1). Accordingly, the new technologies first need to be sold to the innovators and then given time to mature and become acceptable to the pragmatists. The importance of 'champions' is recognized by most vendors, who will tend to suggest that individuals are appointed to become the project team's 'hub' for communications.

As with many change initiatives, it is vital to help individuals identify: 'what's in it for me?' This may be a simple process, perhaps contrasting familiar frustrations with existing methods of project communication (e.g. working on out-dated designs, time wasted in searching for documents, slow or incomplete communications, etc.) with the individual, often intangible, benefits to be gained through electronic collaboration (e.g. no more repetitive form-filling, task simplification, etc. – even a richer job specification, as described in Box 6.3). If scepticism cannot be tackled by 'carrot' benefits, then it can, to some extent, be enforced by a 'stick' approach, perhaps making use of the collaboration technology mandatory for all project-related communications through contracts and protocol documents at the inter-company level, along with individual contracts and job descriptions at the user level.

Box 6.3   New career paths in collaboration management (BIW Technologies, 2005a)

When Mace managed the construction of the Wellcome Trust's new London headquarters, it nominated a document controller who, after initial training by BIW, became the key team member. As new members were inducted to the team, he was able to use the system's reporting tools to monitor their use of the system and identify potential training needs. In each case, he supplied further training or amended the project protocols to encourage adoption of electronic data exchange. In the process, his own role changed from being a conventional 'document controller' associated with a single project to being a capable of managing system implementation, training and consultancy support on two or three projects simultaneously.

Agreeing exchange standards

A key stage is the collective compilation of a protocol document detailing the team's processes, its deliverables and the technologies and information exchange standards it will use. Construct IT's 'How to Manage e-Project Information' (2003) and Project Information Exchange (PIX) Protocol (Building Centre Trust, 2004) have helped fill this gap. The PIX Protocol extends from what IT infrastructure, software applications, document and drawing systems and Internet usage policies end-users currently use individually, to what formats will be used for each type of information and what document naming/numbering conventions will be used. Such attention to detail is important. For instance, agreeing distribution lists can help streamline a project and avoid confusion, targeting information on those who need to know.

Training

As most systems are Web-based, their use of common Website navigation features such as hypertext links and clickable icons will quickly overcome some resistance, and proficient IT users may even start using the systems with minimal training. But it is naïve to think that, however intuitive the vendors claim their systems to be, new users will be able to start using a system effectively without some initial training or support. Appropriate training is vital and needs to be moulded to the needs of different groups of end-users. The introduction of this technology to a project team can also mean new project roles and responsibilities for some end-users.

Cost

The costs can fluctuate depending on how the software is licensed, on who is paying, on the size, complexity and duration of the project or programme (which have a direct bearing on numbers of users, storage, etc.) and what stage it has reached, on training and other implementation requirements, on depth of required functionality, etc. The main distinction is between customers paying a large, up-front perpetual or term licence fee – usually so that they can host the application themselves and based on a certain number of users or 'seats' – or paying smaller fixed monthly subscriptions to rent the software on a remotely hosted Software-as-a-Service (SaaS) basis to use as much as they want (i.e. no limits on the number of users or documents, or on storage capacity) for as long as they need it. Several of the leading construction collaboration providers active in the UK market have opted for the latter model. The precise cost is normally negotiable, but will tend to reflect the size, value and/or scope of the capital project or programme concerned. Some clients, consultants and contractors have moved beyond the 'pay-as-you-go' model and – echoing the partnering model – have signed longer-term contracts with providers, whereby they commit to manage all their projects using that provider's system for, say, a five-year period. From the customer's perspective, such strategic relationships can result in major savings as they can negotiate discounts for large numbers of projects, and integrate the collaboration application more closely into their business processes; the provider, on the other hand, gets greater certainty of future workload and income.

6.6 Moving beyond collaboration

However, these customer relationships may not always be restricted to the continued supply of an existing service; they can extend to the development of new areas of functionality within the main collaboration platform or even to separate, but usually complementary, solutions.

One of the earliest UK examples saw BIW Technologies develop a system for retailer Sainsbury's capable of producing an electronic Health and Safety File, building on information already routinely exchanged using BIW's platform (see Box 6.4).

> **Box 6.4 Electronic Health & Safety File (BIW Technologies, 2005b)**
>
> The Health and Safety File was traditionally a cumbersome and expensive-to-produce library of documents, often filling numerous ring-binders and not completed until weeks after project hand-over. Once compiled, it would be passed to the client, who must then store it, maintain it and make it available for operation and maintenance, major alteration works, or as documentation for future owners. It is, therefore, a key part of a built asset's 'whole-life' documentation.
>
> Sainsbury's had been using BIW and, recognizing the inherent problems of a paper-based File, sought an 'electronic' online solution. Working with BIW, Sainsbury's created a list of health and safety 'attributes' that could be used to electronically 'tag' drawings or other documents when published to the collaboration platform – importantly, a single document could be accessed from different sections of the online File. Instead of being compiled retrospectively at the end of a project, the health and safety coordinator could use the BIW system to start building and safeguarding the integrity the File from the outset.
>
> The system was tested on refurbishment schemes at Huntingdon, Bradford and New Barnet during 2002, and was delivered more quickly and economically at the end of each project's construction. It was then tested on a substantial store extension project in Somerset, but news of the pilot projects stimulated numerous requests from other teams (covering the whole range of Sainsbury's projects) to use the new functionality. In late 2002 Sainsbury's decided to deliver electronic Health and Safety Files as standard for all projects from 1 January 2003. The File is faster, easier and cheaper to compile, maintain and update, and is more accessible to FM staff who need to manage post-construction operation and maintenance processes throughout each built asset's 'whole life'.

Another example of paper-intensive activities within the AEC industry concerns e-tendering. Traditional tendering is a long, complex, paper-driven and – above all – expensive process, but as many organizations were using collaboration systems to manage the processes leading up to the appointment of contractors, subcontractors and suppliers, it made sense, therefore, to use the same system for tendering. Several of the leading construction collaboration providers, therefore, offer e-tendering options (including 4Projects, Asite, BIW and Sarcophagus).

The development of new areas of functionality was also driven by a realization among vendors that creating online document repositories was relatively simple. Even the heavy graphical requirements of most AEC projects do not pose insuperable barriers, so most of the leading construction collaboration technology solutions currently offered in the UK market tend to share many common features. Where they differ is the extent to which they allow users to manage particular AEC project processes online. As a result – and in addition to the health and safety and e-tendering examples already given – there have been various other developments, including:

- Online trading – as well as e-tendering, some of the collaboration vendors (e.g. Causeway, Asite) are looking at managing invoices, purchase orders and remittances through electronic data exchange 'hubs'.
- Project archiving – creating offline archives of all project documentation so that clients and supply chain members retain a complete record of inputs to the project in question.
- Internal communication or knowledge management – some solutions can be configured to function as 'Intranet' products working within a firewall to manage information within an enterprise.
- Process support – while most vendors support various key AEC document processes (RFIs, instructions, change orders, etc.), they vary in their levels of sophistication. Some applications allow teams to exactly replicate paper-based contract processes, perhaps to suit particular contracts (e.g. NEC), creating workflows whereby information is channelled to the team members involved in those processes.
- Cost management – some UK vendors (e.g. Causeway) already have financial solutions as part of their overall product portfolio, but not necessarily integrated with their Web-based solutions. This is already offered in the US market by firms such as Meridian and Primavera, and BIW launched a project cost management solution integrated with its online collaboration platform in late 2006.
- Defects management – in January 2006, BuildOnline (now CTSpace) launched an online mobile data capture service to help firms log defects more accurately, and BIW followed suit in 2007. With browser-based tools now available on mobile devices, users no longer need to be seated at their desks. They can be out on-site, and information can be uploaded wirelessly to the main collaboration platform.

6.7 Conclusions

Much has been written during the past two decades about the 'Internet revolution', but so far as the AEC industry in the United Kingdom is concerned, it is perhaps better to describe the process as 'evolution'. As seen, the AEC sector's conservatism has been a factor in the slow adoption and use of new technologies, and the gradual increase in popularity of construction collaboration technologies has occurred against a background of industry culture changes and of IT advances, particularly since the mid to late 1990s. These changes have not stopped. The industry, the IT tools and the telecommunications at its disposal continue to evolve, and it is unlikely that individuals, organizations and project teams that have embraced such changes will want to revert back to 'the old ways of doing things'. Instead, there may be continued development in at least four key areas:

(1) Declining reliance on paper-based processes to share project information.
(2) Greater integration between design, construction and future operation and maintenance, and between people, processes and technology.
(3) More transparent, longer-term commercial relationships, both within conventional project teams and in relation to technology vendors, and – as a result....
(4) An improved industry reputation for efficiency, reliability, safety and innovation.

References

Autodesk (2003) *Opportunities and Unfinished Business: The Autodesk 2003 Survey of Issues and Trends in the UK Design Community – Focus on Building & Construction*, Autodesk, Farnborough.

Barbour Index (2003) *The Barbour Report 2003: Influencing Clients: The Importance of the Client in Product Selection*, Barbour Index, Windsor.

Becerik, B. (2004) A review on past, present and future of web based project management & collaboration tools and their adoption by the US AEC industry. *International Journal of IT in Architecture, Engineering and Construction*, **2**(3), October, 233–248.

Becerik, B. and Pollalis, S.N. (2006) *Computer aided Collaboration in Managing Construction.* Harvard Design School, Department of Architecture, Design and Technology Report Series 2006-2.

BIW Technologies (2005a) *Mace and Wellcome Trust: Collaborating at the Cornerstone* (BIW case study). Available at <http://212.67.202.17/%7Ebiwtech/cp_root/storage/documents/Wellcome%20Trust%20HQ%20PDF.v1.pdf> (accessed 16 September 2007).

BIW Technologies (2005b) *Producing an Electronic Health and Safety File* (BIW case study). Available at < http://212.67.202.17/%7Ebiwtech/cp_root/storage/documents/Sainsbury%20HandS%20PDF.v3.01.05.pdf> (accessed 16 September 2007).

Building Centre Trust (2004) *PIX Protocol Guide and Toolkit.* Building Centre Trust, London.

Butler Group (2003) *Workgroup and Enterprise Collaboration – Reducing the Costs and Increasing the Value of Collaborative Working*, Butler Group, Hull.

Construct IT (2003) *How to manage e-project information.* Construct IT/ITCBP, Salford.

Croser, J. (2003) Extranet extras. *Architects' Journal*, 20 March, 73–74.
Davies, N. (2004) *CAD Managers' Survey Results: UK Architectural Engineering and Construction Industry*. Cadconsultancy, London. Available at <http://www.cadconsultancy.co.uk/downloads/Cadconsultancy-CADManagersSurveyResults.pdf> (accessed 11 April 2006).
Davies, N. (2006) *CAD Managers' Survey Results: UK Architectural Engineering and Construction Industry*. Evolve Consultancy, London. Available at <http://www.eatyourcad.com/article.php?incat_id=865> (accessed 11 April 2006).
Davis, M. (2003) Earning interest on Knowledge Capital. *Butler Group Review*, June, 21–22.
Department of Trade and Industry (2004) *Business in the Information Age: The International Benchmarking Study 2004*, DTI, London. Available at <http://www2.bah.com/dti2004/pdf/dti-2004_report.pdf> (accessed 11 April 2006).
Egan, J. (1998) *Rethinking Construction, Report of the Construction Task Force*. HMSO, London.
Hansford, M. (2002) Virtual togetherness. *New Civil Engineer*, 5 December, 22–23.
Howard, R. (2002) *A Study of the Productivity Benefits of some Process Changes in the Building Industries of Denmark, Sweden and the UK*, Technical University of Denmark, Denmark.
IT Construction Best Practice Programme (2003) *Using a Project Extranet to Support Partnering in a Prime Contract* (ITCBP case study 041, London: ITCBP. Available at <http://www.itconstructionforum.org.uk/uploadedFiles/041ANDOVER.pdf> (accessed 11 April 2006).
IT Construction Forum (2004) *Survey of IT in Construction: Use, Intentions and Aspirations*. IT Construction Forum, London.
Kalay, Y.E. (1999) *The Future of CAAD: From Computer-Aided Design to Computer-Aided Collaboration*, CAAD Futures '99 Conference, 7–8 June, Atlanta, Georgia.
Latham, M. (1994) *Constructing the Team*. HMSO, London.
Latham, M. (2004) The cynic's bestiary. *Building*, 30 January, 20.
NCCTP (2006) *Proving Collaboration Pays*. NCCTP, London. Available at <http://ncctp.constructingexcellence.org.uk/downloads/making_collaboration_pay.pdf> (accessed 16 September 2007).
Orr, J. (2002) *Keys to Success in Web-Based Project Management: Lessons learned from the Chicago Transit Authority Capital Improvement Program*. Cyon Research.
Ruikar, K., Anumba, C.J. and Carrillo, P.M. (2006) VERDICT – An e-readiness assessment application for construction companies. *Automation in Construction*, **15**, 98–110.
Stratagem/DTI (2003) *e-Business Sectoral Impact Assessment for General Building Contracting within the UK Construction Industry*. Department of Trade and Industry, London.
Verheij, H. and Augenbroe, F. (2001) *A Survey and Ranking of Project Web Site Functionality*. Georgia Institute of Technology, Atlanta.
Wilkinson, P. (2005) *Construction Collaboration Technologies: The Extranet Evolution*. Taylor & Francis, London.

7 Agent-Based Systems: The Competitive Advantage for AEC-Specific e-Business

Esther A. Obonyo and Chimay J. Anumba

7.1 Introduction

With the e-business world growing at an unprecedented pace, new market realities and challenges are emerging. One of the main problems in the digital world is the sheer volume of information available in heterogeneous format making the retrieval and processing of information for electronic trade a tedious and laborious process. This problem is evident in AEC-specific e-business. Though there have been a number of portals in this sector enjoying varying degrees of success, a lot of information is still processed by human agents. The successful implementation of an optimal system has been impeded by the existence of semi-structured or non-structured information held in various formats. This problem is further compounded by the general challenges of application integration. Consequently, a significant amount of time is spent by designers in gathering relevant information. This paper is based on a prototype system developed to demonstrate how agent technology can make AEC-specific e-business more effective and more efficient. The paper introduces the selected domain, pointing out the challenges in executing online activities. An overview of agent technology and the vision for agent-based e-business have been given. The paper also describes the conceptual design for the prototype system and the implemented architecture for the final prototype system. It ends with a few concluding remarks reaffirming the need to incorporate agents AEC-specific e-business and also points out the need to develop realistic expectations when deploying agent-based systems in commercial settings.

7.2 The current context

The AEC (Architecture, Engineering and Construction) business is experiencing an era of dynamic, highly competitive markets that continually place new and more demanding requirements on its operations. During this era, the UK's AEC sector in particular has experienced acute pressure to modernize and improve its operations. Such concerns were best articulated in the Egan Report (Egan, 1998), which highlighted the main shortcomings of operations within the built environment. A significant problem area was significant levels of waste and rework that can be attributed to fragmentation and discontinuity of teams within the industry. The Egan Report specifically cited the use of new technologies in the design of building and components and in the exchange of design information as a potential solution.

These issues highlighted above are not peculiar to the AEC sector and there are also many similarities in the solutions being explored. One of the key technologies used to enhance team working is the Internet. Since the 1990s many organizations in the various sectors have embraced the use of the Internet in their business processes convinced that Internet-based communication technology would decentralize organizations by making information accessible to geographically distributed work teams, suppliers and partners. The evolving collaboration challenges are being matched by new Internet-enabled technologies and delivery mechanisms, which emphasize on flexibility and scalability.

Many leading researchers have suggested that the Internet, by supporting communication throughout the development process and across the different parties, holds the key to an integrated, and therefore a more productive, AEC sector. Existing literature is strewn with evidence demonstrating the potential of Internet-based collaboration for this industry (Taylor and Björnsson, 1999; Björnsson, 2000a, b; Rakow, 2000; Vikkula, 2000).

The AEC sector's ability to significantly exploit the Internet's potentials depends directly on the existence of a central repository that creates a more versatile framework for capturing, storing and reusing information throughout a built asset's life cycle. Such a repository connects players, projects and products of the construction process and its value lies in its ability to present information on the built asset as a detailed report comprising of, for example, CAD drawings and full product specifications.

The format used in storage must also mirror the various players' information consumption behaviour. Designers should, for example, be able to seamlessly pick and place product specifications from manufacturers' predefined catalogues and define them in their own models. The data warehouse must provide interfaces that allow it to be operated on by panoply of analytic applications, such as CAD, estimating and scheduling tools, energy performance simulators and code compliance software.

The specification and procurement of construction products within such a system becomes a simple task and information is exported across applications without significant investments in new infrastructure. With such a system in place, digitally held information on construction products is easily used throughout the construction phase in:

- Servicing the design information needs.
- Predicting and managing the sequencing and scheduling of building products in accordance with a master project timetable.
- Storing material and equipment attributes for use during building operations or facility management phase.

After a rigorous review of trends in Internet-enabled collaboration for the construction industry, Laiserin (2001) established that although existing systems have significantly exploited the Internet's abilities to transcend space, time and competition, there is a large scope for improvement. Current information architectures and systems already meet many requirements such as availability of digital information on construction products. During the past 10 years or so, the current electronic information age has experienced significant developments. The increase in inter-networked environments has led to more electronic information becoming available to people. Yet networks of information present the electronic information community with a fundamental problem: digitally held information exists in incompatible formats within heterogeneous repositories. Subsequently, systems holding digital information have become 'digital islands'.

The problem is further exacerbated by the vast quantities of information that are available in these 'digital islands'. The total time spent locating and filtering through irrelevant information is quite substantial. There is a greater need for tools that help users to manage and navigate through the maze of digitally held information. Such tools must also be capable of integrating and communicating with disparate information repositories to minimize the effort spent re-keying in data.

The increase in awareness of the potential digital information has led to the introduction of a wide range of services, such as the various product libraries, attempting to manage the substantial volume information that is becoming available. However, due to such services being heterogeneous and lacking in extensibility in their inherent functions, they further perpetuate the fragmentation within the construction industry. There is still a significant amount of effort expended in searching for relevant information on construction products, and when the information is finally located, reuse in a different application would require a user to re-key in data.

All these issues were quite correctly diagnosed by a group of researchers as the 'digital anarchy' problem (Radeke, 1999). Ugwu *et al.* (2002) affirmed the existence of such problems within the electronic procurement of construction products and highlighted the potential of software agents.

7.3 Understanding agent-based systems

There are many divergent views on the exact definition of software agents. It is therefore necessary to develop a working definition for this chapter. The concept of software agents within this context has been explored from the viewpoint of leading researchers, who define software agents using their general characteristics. Brustoloni (1991), Ferber (1999), FIPA Architecture Board (2001), Jennings *et al.* (1998), Jennings and

Wooldridge (1998), Lieberman (1997) and Maes (1994) have defined software agents using the following characteristics:

- Software agents exist in an environment.
- They can sense the conditions in the environment and such senses may affect how they act in future.
- Software agents are adaptive and capable of learning.
- They are proactive, exhibiting goal-directed behaviour.
- They execute their tasks autonomously (i.e. without human intervention).

The adopted definition based on a synthesis of these characteristics is: 'Agents are systems capable of autonomous, purposeful action in the real world.'

A single classification structure for agents also remains an unresolved issue. Examples of different structures include the simple typology provided by Nwana (1998) and extended by Ugwu *et al.* (1999). Ferber (1999) also provided a classification system based on anticipation and reaction. The reviewed literature revealed that there are so many different structures for categorizing agents that it is not possible to show all the different agent types in one diagram. The most holistic approach defines the primary base of all agents and allows users to create extensions from the base depending on the application domain. Franklin and Graesser (1996) proposed a taxonomy of agents analogous to the biological classification system based on this approach.

This taxonomy places autonomous agents at the pinnacle of a tree structure and defines software agents as a sub-class of computational agents, which in turn are a sub-class of autonomous agents. Extensions from software agents can be made on a number of schemes (Brustoloni, 1991; Franklin and Graesser, 1996; Jennings and Wooldridge, 1998; Gini, 1999;), for example tasks performed and applications.

7.4 A roadmap of agent-based systems in e-business

The vision for agent-based systems in e-business, as shown in Figure 7.1 defines a scenario in which all parties automate trade, negotiation and collaboration in electronic market places via agents. The use of buyer agents, seller agents, middle agents embedded on servers, desktops, mobile and handheld computers, mobile phones is expected to result in an increased volume of deals and a reduction of transaction costs (Shehory, 2003).

The potential contribution of software agents to the success of e-business initiatives lies in their ability to collect Web content from diverse sources, process the information and exchange the results with other programs

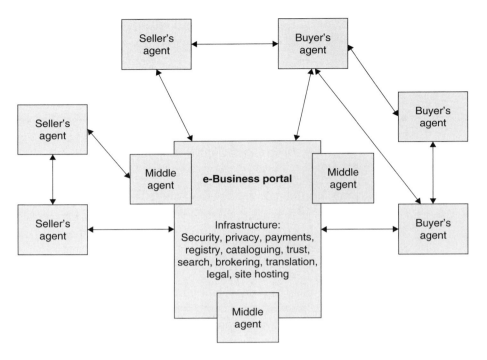

Figure 7.1 A vision for agent-based e-business

(Berners-Lee *et al.*, 2001). Blake and Gini (2002), Blake (2002) and Samtani (2003) provide further information on the rationale behind the use of agent-based systems in business operations in the digital era. The approach adopted in the development of the prototype described in a subsequent section is based on the premise that the Semantic Web is evolving and the paradigm of software agents is gradually becoming an indispensable technology for companies in the modern era.

The expansion in the use of the Web and its exponential growth are well-known facts. However, a problem has emerged from the very core of the success of the Web. Available statistics quantify the significant problems related to the use of the Web in its present form. The Web has evolved into a very large, unstructured but ubiquitous database. It was estimated that just textual data on the Web is in the order of one terabyte (Baeza-Yates, 1998). Such information overload on the Web has resulted in substantial losses. KPMG (2000) established that two-thirds of 423 European and US organizations assessed in one survey suffered information overload and failed to share knowledge. It has also been established that on average, US employees spend eight hours a week (16% of their working week) searching for and processing external information. This translates into a financial cost of $107 billion a year (with each employee costing their firm $10000 annually). Such substantial financial loses underscore the need for change.

The problems related to the use of the Web can also be assessed at the micro-level. During the past decade many organizations have invested in various components of information systems to support their business functions. There were great expectations that the use of Internet-enabled systems would improve productivity and reduce cycle times in various operations by collecting and providing the right information. A significant proportion of the forecasted gains has not been realized partly due to the lack of inter-operability between systems (Madhusudan, 2001). An extensive study conducted at the Center for Research in Electronic Commerce (CREC), at the University of Texas-Austin to assess the collaborative capabilities of companies also arrived at the same conclusion (Samtani, 2003). Collaboration has subsequently become more important for any company seeking to make the most out of e-business (Deloitte Research, 2002).

Clearly, there is a need for Internet-based technologies that address the availability of information and the ability to seamlessly exchange and process it across different applications in different organizational units. However, it is important to note that agent-based e-business applications are still in the early stages of development and leading researchers suggest that there is likely to be a time lag of three to five years between research prototypes and real-life applications (Luck *et al.*, 2003).

7.5 APRON: An agent-based prototype system for AEC-specific e-business

APRON (agent-based specification and procurement of construction products) project emerged from a partnership between Loughborough University and BIW Technologies. BIW, a construction-industry specific e-business portal, was launched towards the end of 1994 following encouragement by the UK Government for increased Internet use in construction. The original BIW portal Website was launched at the beginning of 1995. Its primary feature was a free, searchable directory of building products, manufacturers, suppliers, contractors and other industry professionals. The company's core product was the BIW Information Channel, an innovative Web-based communications platform for the AEC sector that sought to eradicate such problems as inaccurate, inadequate or inconsistent information in a project from inception through to facility management through creating a central repository of information about the built asset. It allowed users to access information exactly tailored to their needs. However, at the inception of APRON, the system lacked the ability to retrieve and manipulate information on construction products from the various manufacturers in an efficient manner.

The specification and procurement of construction products challenges the frontiers of computing due to the following peculiarities and complexity it inherits from the AEC sector as a whole:

- The existence of many different disciplines in a single project.
- Physical dispersion of the project team.
- The use of many different computing applications.
- The time-related fragmentation (possible discontinuities across inception, design, construction and exploitation phases).

In the 1990s, there was a prevalent view that the AEC sector was characterized by inertia as far as embracing new technology was concerned (Egan, 1998). The success experienced by BIW indicated that this view was slowly losing validity. Through a series of discussions with manufacturers, BIW had established that the market for building and construction products had become more open and competitive, particularly with the widespread adoption of Internet-based technologies. Manufacturers were free to move around the globe with their offerings and they competed as well as co-operated with each other. They needed to introduce new information on construction products quickly and required systems that would allow them to do so. Such changes in requirements resulted in doubts on the adequacy of the existing Internet-held information on construction products. Exploratory work by the company had resulted in LISP-based components within AutoCAD that interfaced with attribute-related information from selected manufacturers. The deployment of this initiative into a real-life application was challenged by the 'digital anarchy' problem. There was a need for a system that could automate extraction and structuring of information. The system also needed to handle heterogeneity in the display mechanisms used by product manufacturers.

APRON was conceived as a sub-project of work that was already in progress within an initiative led by one of the founding members of the company who was an Architect by background. This initiative focused on developing tools that processed information from manufacturers' catalogues and was strongly rooted in work that had been done by Visual Technologies in collaboration with BRE under the ARROW (Advanced Reusable Reliable Object Warehouse) initiative (Amor and Newnham, 1999).

ARROW provided an Internet-based framework facilitating the identification of manufactured products meeting specific design criteria. This framework provided manufacturers with toolkits for mapping their product information into formats that could be utilized by ARROW. An AutoCAD demonstrator, which identified the closest matching manufactured product and imported the result as a DWG into an AutoCAD drawing, was implemented. Visual Technologies provided the technology for storing object information within the drawing. The import functionality was developed using AutoLisp script and Java classes.

The research team at BIW had done further work on ARROW's AutoCAD-based models, which they now referred to as I-Components (intelligent components). A key outstanding issue was establishing a mechanism for gaining structured data from manufacturers. BIW's I-Components were modelled after the ARROW approach, which was largely dependent on convincing manufacturers to publish information in a new style.

BIW realized that an ideal solution had to automate the tasks inherent retrieval and the company started reviewing the trends in Internet-based information management. From the mid-1990s, researchers presented the concept of intelligent (software) agents as a panacea for Internet-related complexities. BIW quickly identified agent technology as a possible solution to their information retrieval challenges. The research team commissioned APRON as a project to investigate the potential of using agent technology in automating the retrieval of information from product manufacturers.

The context was restricted to accessing product information from AutoCAD. Issues related to defining the parameters for different products were deliberately excluded from the research. Though such issues are significant, there would not be an ideal test bed for new agent application (EURESCOM, 2000). Such issues have inherent complexities that would require the deployment of highly sophisticated user interface agents: a task that was perceived to be too challenging to be an objective of a pilot project/proof of concept research. Developing agents to support the collection of knowledge, and therefore addressing the information overload problem as outlined in Chapter 1, must precede the refinement of context-aware agents that can diffuse information based on individual product characteristics.

The scope of the problem was refined through an evaluation of the existing Web-based portals supplying information to support the specification and procurement of construction products. The McGraw Hill Sweets Product Library (http://sweets.construction.com/default.jsp) was selected for use in a case study conducted to do just that. Its qualification for study was based on its influence and representation. The library has an influence in the AEC sector in terms of scale and popularity. It also has a structured and comprehensive database of product manufacturers under various categories.

The central question in the assessment of the repository was establishing its performance in supporting seamless information flow in the specification and procurement of construction products. The assessment revealed a number of problems related to the content of the library and collaboration support. From the home page, one has to manually follow links to the Products Centre from which they can either use a search engine or browse to view a list various companies offering a desired product as depicted in Figure 7.2. This company listing provides links to Websites hosted by different companies. It does not have links to the actual information on existing products.

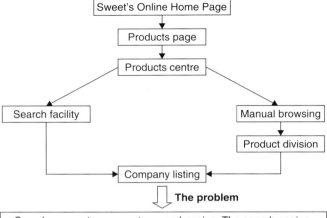

Figure 7.2 **The existing digital sources**

In an exemplary scenario, lighting bulbs were specified as the search parameter. The search facility failed to recognize the specified words as acceptable terms. The navigation style was then switched to manual browsing and although the desired information on product specifications was finally accessed, the process was serial and tedious. It involved browsing through 111 companies listed under the electrical section and manually following links to various product listings for each of these companies. Navigation in this digital space is quite similar to searching paper-based catalogues. This diminishes the value of going digital and the process has failed to eradicate the paper-based communication patterns and sequential work methods that depend on paper. All these problems are summarized in Figure 7.2.

As far as collaboration support was concerned, the Library failed to address the fragmentation and the discontinuity of teams within the industry. True collaboration support would require a system that makes it easier for multiple participants to share information interactively and asynchronously. Sweet's Library provides links to Web pages holding documents posted by various manufacturers; it does not provide a mechanism for making the information in the documents more accessible across the building team.

Some of the links in the evaluated portal lead to CAD drawings that could be exported to end-users' applications. However, these were generally standalone CAD drawings and models, which did not represent the building products as both objects and elements. Such drawings and models lack interfaces to the rest of the information on product specification in the product catalogues. They cannot be readily passed on to a different user without an element of re-keying of data.

Clearly, there is still a need to improve things further and make the use of Internet more versatile as far as capturing, storing and reusing information throughout a built facility's life cycle is concerned. The domain analysis revealed that existing systems lack the ability to define the components of the final product as a detailed report that could be used across the various construction phases by the various disciplines regardless of their IT systems.

The results of the domain analysis established that despite the massive increase in the availability of Internet-based information on construction products, the potential gains in productivity remain unrealized. This can be attributed to the absence of effective structures for managing information. Unfortunately, as more information becomes available, the potential seriousness of this problem also increases. These problems have been attributed to the factors outlined in Table 7.1.

The contention of this chapter is: tackling the problems cited above is of strategic importance in the specification and procurement of construction products. The UK's AEC sector has been under great pressure to radically improve its performance and a good framework for managing Internet-held

Table 7.1 The factors behind the problems

Factor	Description
Volume	With at least 1 tetra byte of information available on the Internet, end-users experience difficulties in absorbing, filtering and analysing data.
Understanding	End-users cannot anticipate and plan solutions for any possible information complexities.
Flexibility	End-users must adapt to the different requirements when dealing with information from multiple sources – each manufacturer has a unique display strategy.
Information detection	The existing systems lack an efficient search mechanism and do not provide end-users with any means of determining the existence of relevant information.
Access	The existing systems bring back documents rather than the desired information on construction products.
Relevance	Because the existing systems lack efficient search mechanisms, the end-users are unable to filter out relevant information.
Validation	There is no efficient mechanism of establishing the accuracy and quality of the obtained information.
Monitoring	The existing approaches lack long-term coverage allowing proactive update detection

information is imperative if the sector is to successfully respond to this challenge. Moreover, with the Internet transcending physical boundaries, the market in which construction firms operation has changed and is continuously evolving. A good system of managing Internet-held information would provide leverage and competitive advantage in this context.

7.6 APRON's conceptual design

Figure 7.3 depicts a hierarchical view of goals to be achieved by the system. In order to achieve these functions, three categories of APRON agent roles will be created. These roles and their respective tasks and protocols have been depicted in Figure 7.4. The enumerated labels in the roles, for example 1.1.1, refer to the goals shown in Figure 7.3.

Figure 7.5 shows an agent template depicting how the three agent categories communicate with one another. The blocks represent agents that serve the roles previously depicted in Figure 7.4 Inter-agent communication between the agents is represented by conversation18_1 to conversation27_1.

These conversations tally with the protocols previously given in Figure 7.5 for the executions of various tasks and roles. Table 7.2 matches these protocols with the corresponding conversation(s). The conversations have not been enumerated in any order.

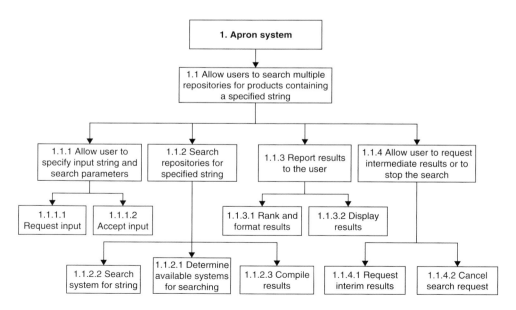

Figure 7.3   Hierarchical view of agent goals

Agent-Based Systems 115

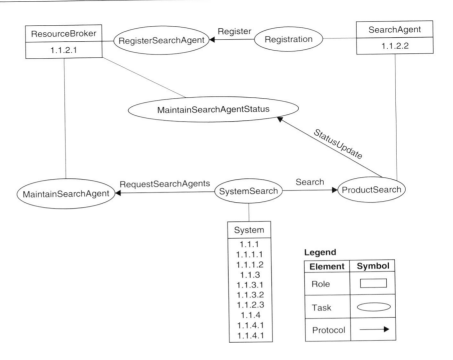

Figure 7.4 Agent Roles Tasks and Protocols

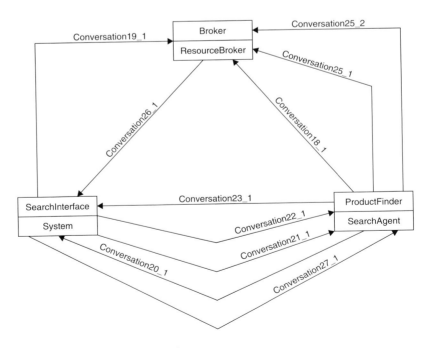

Figure 7.5 Agent communication

Table 7.2 Agent communication

Agent	Role	Task	Conversation
Broker	ResourceBroker	RegisterSearchAgent ManageSearchAgent MaintainSearchAgent Status	18_1 19_1 & 26_1 25_1
SearchInterface	System	SystemSearch	20_1, 21_1, 22_1, 19_1, 26_1, 23_1 & 27_1
ProductFinder	SearchAgent	Registration ProductSearch	18_1 20_1, 21_1, 22_1, 23_1, 25_1, 25_2 & 27_1.

7.7 The implemented APRON architecture

The implemented APRON prototype was modelled using the established mediation/wrapper methodology that was used in, for example, the InfoSleuth prototype (Bayardo *et al.*, 1996) and the SEEK prototype (O'Brien *et al.*, 2002).

In this approach, the developed solution comprises a software middle layer between the semi-structured repositories of construction products and the end-user applications utilized in specification and procurement. This is depicted in Figure 7.6.

There are three distinct, intercommunicating layers. The construction product manufacturers occupy the top level in the architecture. They display details of the various product offerings as semi-structured data on the Internet. The kernel of the architecture is the e-marketplace of an information provider, who uses the Internet to ensure that requisite project information is available to all the key players in the construction project supply chain.

The e-marketplace hosts the APRON solution, which consists of the Download, Extraction, Structuring, Database, Search and Procurement Modules. The e-marketplace also hosts a repository of relevant standards that can be used as XML schemas for structuring the extracted information. The final layer comprises the end-user firms in the specification and procurement of construction products. The focus in these firms is providing an automated interface to the computing applications used by specifiers and procurers. A client application has been deployed to allow such end-users to communicate with the APRON Web Service.

The APRON system provides a link between the Website holding product information and the applications used in the specification and procurement of construction products. The Download Module maintains real-time access with these Websites. Text is then extracted from the downloaded file, and using previously defined XML schemas, structured into a

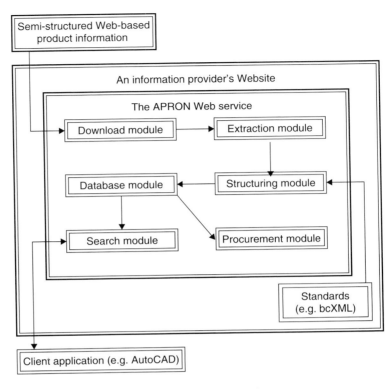

Figure 7.6 The implemented APRON prototype system

context-specific format. Industry standards, such as ifcXML and bcXML, can be easily adopted and used to create an XML template for structuring the extracted information. The APRON system also stores the relevant information in a database. Information for the specification and procurement of construction products is obtained from this database. Specifically, APRON, provides a Web-based search engine, which can be used to execute context-specific queries for construction products. The APRON solution also offers a framework for automating the procurement of specified products. The Procurement Module has two types of agents: a buyer agent and a seller agent representing product procurers and product sellers, respectively. The two agents exchange requisite information and automate the transactions involved in the procurement of construction products. Obonyo et al. (2004) provide a more detailed description of the APRON prototype.

7.8 Discussion and conclusions

There is a need for Internet-based technologies that address the availability of information, the ability to exchange it seamlessly and the ability to process it across different applications in different organizational units. The chapter has outlined the competitive advantage of using agent technology in AEC-specific e-business. The deployment of APRON, a prototype system for the specification and procurement of construction products confirmed this leverage. However, it is important to note that the deployment of such a system in commercial settings will not be an easy task.

The greatest challenges probably emanate from the issues inherent in e-business as an application domain. A review of existing commercial applications revealed that agent-based e-business applications are not as advanced as applications in domains such as telecommunications. Shehory (2003) pointed out that the results of recent surveys indicate that the adoption of agent technology in e-business has been very slow. One of the primary factors hindering rapid adoption of agent technology has been the global economic trend towards a recession.

The early years of the 21st century were characterized by companies postponing their major IT investments and becoming more risk averse and savvier in their IT investments. Web-based technologies, in particular, were adversely affected following the end of the Internet euphoria: an era that peaked between 1999 and 2001 and was characterized by 'overspending' (Benjamin *et al.*, 2003). However, this problem will diminish in significance with the passage of time. In the long run, the economic cycle will tend towards prosperity and IT investment will increase.

The slow uptake of agent-based e-business applications could also be attributed to 'failures' among researchers in the agent community. There was excessive and somewhat misleading enthusiasm among agent technology pundits in the late 1990s (Luck *et al.*, 2003). This resulted in unrealistic expectations and disappointments when the promised benefits were not fully realized. Agent technology does not hold the ultimate answer to the 'digital anarchy' problem: it contributes to the solution but it must be applied in conjunction with other building blocks such as RDF, ontologies, XML and Web services. AEC-specific agent-based applications must be viewed from this perspective.

The problem related to unrealistic expectations and resulting disappointments may have been compounded by the time lag between research activities and commercial implementation. In general, research prototypes in agent technology do not immediately result in commercial applications. A large-scale adoption of the APRON concept will not be an exception to this general rule.

Prospective investors in agent technology must therefore be guided into developing modest expectations. This can be achieved through educational workshops and courses in various aspects of agent technology. Agentcities (http://www.agentcities.org/) and AgentLink (http://www.agentlink.org/) have championed awareness-creation in the European context. The two groups have promoted awareness in agent technology in universities and companies across Europe, through, for example, awarding deployment grants for commercial agent-related projects. AEC-specific projects were conspicuously missing from the ventures supported by the two organizations. Agent-based e-business applications have been generally executed as small projects with varying degrees of success. However, such 'individual' efforts address a small fraction of the problem and are highly unlikely to produce solutions effectively covering all areas (Willmott et al., 2003).

It is imperative for researchers in agent technology within the AEC sector to align themselves with synergies such as AgentLink and Agentcities. Such initiatives provide an opportunity to form interdisciplinary teams that would allow them to exploit the skills and competencies that have already been established by experts in agent technology. The sector can also benefit from events organized within these synergies to promote awareness within the commercial settings. Such events are aimed at getting top-level commercial managers to appreciate what software agents are and what they are capable of. This empowers them to identify the existing opportunities for agents in their mainstream business processes.

Phillips (1999) published a Masters thesis at the MIT with the title *If It Works, It's Not AI: A Commercial Look at Artificial Intelligence Startups*. This thesis analysed the successes and failures of AI technology firms in bringing their products to market and creating lasting businesses. There was a high level of hype in the form of mainstream press and corporate excitement prior to the downfall of AI in the 1980s. The pitfalls that many of these firms fell into included:

- Management inexperience and academic bias.
- Business models, which confused products and consulting.
- Misunderstanding of the target market.
- Failing to manage customer and press expectations.

During the 1990s AI made a subtle comeback in the form of agent technology. Surprisingly, software agents have gone through a life cycle that is not too dissimilar to that of AI in the 1980s. This is evident from Janca's (1995) false prophecy on pervasiveness of agents by 2000. The efforts towards commercialization of AI in the 1980s, particularly the experiences of firms that faltered, should have provided some warnings and suggestions for any company trying to build a business packaging agent technology. Based on the experiences of early commercial applications of agent-based systems, it would appear that history has repeated itself: champions of agent technology repeated the same mistakes made by the AI predecessors in the 1980s. As late as 2003, there were few successful commercial applications of agent technology and many people regarded agents with a heightened mistrust as predictions failed to materialize.

Such problems should not be interpreted as proof of inappropriateness of agent technology. They merely indicate that agent technology is not an industry in and of itself. The reality is: not all entities in a software system require sophisticated capabilities such as reasoning, learning and autonomy and a number of passive components cannot be modelled and used as agents. Agents must therefore be viewed as components enriching larger software systems. In this regard, e-business agent-based applications such as APRON must be developed within the context of Semantic Web Technology. Such pragmatism combined with economic suasion based on savings from automation of tedious, routine and repetitive tasks are indispensable ingredients in developing a strategy that would ultimately cast business managers in the role of allies.

Agent technology holds a real promise but the nettlesome task of res-earchers is to discover how to organize its strength into compelling power. Although the long-term vision should still remain delivering e-business agent-based applications that use sophisticated reasoning, learning or planning techniques, the experience of the 1990s have proved that too much focus in such issues would result in premature enthusiasm. The strength of agents in supporting Internet-based is really using agents to address information overload. The short-term efforts should really focus on information retrieval and processing before building in profiling and learning capabilities.

References

Amor, R. and Newnham, L. (1999) CAD interfaces to the ARROW manufactured product server. In: *Proceedings of the Eighth International Conference on Computer Aided Architectural Design Futures*, Kluwer Academic Publishers, Atlanta, USA, pp. 1–11.

Baeza-Yates, R.A. (1998) Searching the World Wide Web: Challenges and partial. In: *Proceedings of the 6th Ibero-American Conference on AI: Progress in Artificial Intelligence – IBERAMIA 98*, (H. Coelho, ed.), *Lecture Notes in Computer Science*, Vol. 1484, Springer-Verlag, London, pp. 39–51.

Bayardo, R., Bohrer, W., Brice, R., Cichocki, A., Fowler, G., Helal, A., Kashyap, V., Ksiezyk, T., Martin, G., Nodine, M., Rashid, M., Rusinkiewicz, M., Shea, R., Unnikrishnan, C., Unruh, A. and Woelk, D. (1996) *Semantic Integration of Information in Open and Dynamic Environments*, TMCC Technical Report, MCC-INSL-088-96.

Benjamins, V.R., Contreras, J. and Prieto, J.A. (2003) Agents and the semantic web. *AgentLink Newsletter*, An AgentLink Publication. Available at www.agentlink.org/newsletter/ (accessed on 13th Jan 2004).

Berners-Lee, T., Hendler, J. and Lassila, O. (2001) The semantic web. *Scientific American*, May. Available at <http://www.sciam.com> (accessed on 15 January 2002).

Björnsson, H. (2000a) *Impacts of Technology on the A/E/C Industry in the 21st Century*. The Reading Construction Forum, London.

Björnsson, H. (2000b) *eBusiness in Construction*, Presentation at CIFE in Finland.

Blake, M.B. (2002) AAAI-2002 Workshops: Agent-based technologies for B2B electronic commerce. *AI Magazine*, **23**(4), AAAI Press, 113–114.

Blake, M.B. and Gini, M. (2002) Guest editorial: Agent-based approaches to B2B electronic commerce. *International Journal on Electronic Commerce*, **7**(1), 5–6.

Brustoloni, J.C. (1991) *Autonomous Agents: Characterization and Requirements*, Carnegie Mellon Technical Report CMU-CS-91-204, Carnegie Mellon University, Pittsburgh.

DETR (1998) *Rethinking Construction – The Egan Report*. UK's Department of the Environment, Transport and the Regions.

EURESCOM (2000) Model and guidelines for assessing the suitability of agent technology adapted from Project P907-GI, MESSAGE: Methodology for Engineering Systems of Software Agents, Deliverable 1. Available at http://www.eurescom.de/~pub-deliverables/P900-series/P907/D1/P907D1.doc (accessed 22 September 2001).

Ferber, J. (1999) *Multi-Agent System: An Introduction to Distributed Artificial Intelligence*, Addison-Wesley, An imprint of Pearson Education, UK.

FIPA Architecture Board (2001) *FIPA Agent Software Integration Specification*. Available at <http://www.fipa.org> (accessed 30 September 2002).

Franklin, S. and Graesser, A. (1996) Is it an agent, or just a program? A taxonomy for autonomous agents. In *Intelligent Agents III, Agent Theories, Architectures and Languages (ATAL)* (Muller, J.P., Wooldridge, M.J. and Jennings, N. eds). Springer-Verlag, Berlin, Germany, pp. 21–35.

Gini, M. (1999) *Agent and other 'Intelligent Software' for e-Commerce*, Talk presented at the Department of Computer Science and Engineering, University of Minnesota, CSOM, 12 February. Available at http://www-users.cs.umn.edu/~gini/csom.html (accessed 30 October 1999).

Janca, P.C. (1995) Pragmatic application of information agents, *BIS Strategic Report*.

Jennings, N.R. and Wooldridge, M. (1998) Applications of intelligent agents. In *Agent Technology: Foundations, Applications, and Markets* (N.R. Jennings and M. Wooldridge eds). Springer Verlag, Berlin, pp. 3–28.

Jennings, N.R., Sycara, K. and Wooldridge, M. (1998) A roadmap of agent research and development. *Autonomous Agents and Multi-Agent Systems*, **1**(1), 7–38.

KPMG Consulting (2000) *Knowledge Management Research Report*. Available at <http://www.kmadvantage.com/docs/km_articles/KPMG_KM_Research_Report_2000.pdfH> (accessed 3 July 2003).

Laiserin, J. (2001) Liberte, Egalite, Fraternite, Online Collaboration, *CADENCE Magazine*, November.

Lieberman, H. (1997) *Autonomous 1nterface Agents*, Presented at the ACM Conference on Human-Computer Interface [CHI-97], Atlanta.

Luck, M, McBurney, P. and Preist, C. (2003) Agent technology: Enabling next generation computing. *A Roadmap for Agent-Based Computing, Version 1.0, AgentLink Report*. <Available at http://www.agentlink.org/> (accessed 8 November 2003).

Maes, P. (1994) Modeling adaptive autonomous agents. In *Artificial Life Journal* (C. Langton ed.), **1**(1 & 2), MIT Press, 135–162.

Madhusudan, T. (2001) *Enterprise Application Integration – An Agent-Based Approach*, Presented at the IJCAI Workshop on AI and Manufacturing, Seattle, WA, August.

Nwana, H., Rosenschein, J., Sandholm, T, Sierra, C., Maes, P. and Guttman, R. (1998) Agent-mediated electronic commerce: Issues, challenges, and some Viewpoints. In: *Proceedings of the Second International Conference on Autonomous Agents (Agents '98)*, ACM, pp. 186–196.

Radeke, E., ed (1999) *Final Report, GENIAL Global Engineering Networking Intelligent Access Libraries*.

Ugwu, O.O., Kamara, J.M., Anumba C.J. and Leonard, D. (2002) Electronic Procurement of Construction Products, *International Journal of Service and Technology and Management*, Vol. 3, No. 2, pp. 222–237.

O'Brien, W., Issa, R., Hammer, J., Schmalz, M., Geunes, J. and Bai. S. (2002) SEEK: Accomplishing enterprise information integration across heterogeneous sources, *ITCON-Electronic Journal of Information Technology in Construction, Special Edition on Knowledge Management*, **7**, 101–124

Phillips, E.M. (1999) *If It Works, It's Not AI: A Commercial Look At Artificial Intelligence Startups*, Master Thesis, MIT.

Rakow, B. (2000) 21st Century technology: The time is now to embrace new tools for the industry. *Constructech*, **3**(1). Available at http://www.specialty-pub.com/constructech/(accessed on 30th July, 2003).

Samtani, G. (2003) B2B *integration, A Practical Guide to Collaborative E-Commerce*. World Scientific Pub Co Inc, USA.

Shehory, O. (2003) Agent-based systems: Do they provide a competitive advantage? *AgentLink Newsletter*, An AgentLink Publication.

Taylor, J. and Björnsson, H. (1999) Construction supply chain improvements through internet pooled procurement In: *Proceedings of IGLC-7*, Berkeley, CA, 26–28 July, pp. 207–217.

Ugwu, O.O, Anumba, C.J., Nehnham, L. and Thorpe, A. (1999) *Applications of Distributed Artificial Intelligence in the Construction Industry, Research Report No. ADLIB/ 01*, Department of Civil and Building Engineering, Loughborough University and Building Research Establishment, pp. 10–18.

Vikkula, M. (2000) *eBusiness – Current Status and Future Expectations*, Presentation at the eBusiness Seminar September 6th, VTT, Finland.

http://sweets.construction.com/default.jsp.

http://www.agentcities.org/ (accessed 14 August 2002).

http://www.agentlink.org/ (accessed on 20 July 2002).

Willmott, S., Bonnefoy, D., Thompson, S., Constantinescu, I., Dale, J. and Zhang, T. (2003) Towards new generations of Agile Networked Software, *Agentcities Research Report*.

8 The Role of e-Hubs in e-Commerce

Zhaomin Ren, Chimay J. Anumba and Tarek M. Hassan

8.1 Introduction

In the recent past, e-Hubs have been flourishing, either in the form of public marketplaces or as private exchange platforms within enterprises. e-Hubs have been used in a wide range of industry sectors (e.g. Shipping and Trucking, Warehousing, Perishable Goods, Chemicals, Travel, Entertainment, Real Estate, Insurance Services, Manufacturing, Financial Services, and Media). e-Hubs have clearly demonstrated their power as a real-time, global distribution marketplace by dissolving the constraints of time and geography and made it possible for buyers, Business-to-Business and Business-to-Consumers to become more fully engaged. Below are some of the examples:

- According to the ITNET (URL1), the e-procurement contract using Infobank's in-Trade software is worth up to £1.6 million over three years.
- Biotech analytics has won orders worth over £500 000 since the e-Hub went live on 27 November 2000.
- Companies like Ariba, Chemdex, Commerce One, Free-Markets, Internet Capital Group, and SciQuest.com have gained large amount of capital from stock markets.
- e-Steel.com (URL2) serves the $700 billion global steel industry. Its e-commerce solution Website unites buyers and sellers and provides an array of resources for sales and purchasing professionals throughout the steel value chain. Nearly 3000 companies, from 90 countries have become members of e-STEEL, representing all of the key segments of the steel value chain.

e-Hubs create value by aggregating buyers and sellers, creating marketplace liquidity, and reducing transaction costs and time. e-Hubs are changing traditional business transactions and the relationships between enterprises by integrating the supply and demand of an enterprise with its customers, suppliers and partners, and automating business transactions and information sharing. They create unprecedented levels of market transparency and lower the cost of transactions. Furthermore, industry users are able to liquidate their service products, establish exchange networks, and build collaborations through e-Hubs. Companies use e-Hubs to synchronize operations with their demand and supply chains to change product lifecycle.

By bringing a large number of suppliers and customers together and facilitating transaction and collaboration between these players, e-Hubs are able to:

- Configure a multi-channel sell-side solution. It enables a company to market and sell more inventories to existing customers and to establish new customer relationships by connecting to e-marketplaces and portals. e-Hubs allow companies to control and manage all of its sales channels from a single interface.
- Reduce costs significantly due to the enhanced contract pricing leverages during supplier negotiations, transactions, and product development such as reducing paperwork, travelling costs, and inventory costs, allowing multiple sales, and integrating and consolidating payment.
- Increase market transparency as customers and suppliers can track and monitor the whole transaction process, profile their purchases and customers and administration of contracts and authorizations.
- Facilitate the supply chains among enterprises to speed time-to-market through collaboration with suppliers and customers. Steps required for design, execution, prototyping, and changes become workflows in which information is shared seamlessly with partners to create the right design in a shorter period with a quicker ramp to high-volume manufacturing.
- Coordinate logistics and capital management with the business and production process.

8.2 e-Hub concept

e-Hubs have been continuously being studied over the past two decades. The understandings to e-Hubs have also been changing with the quickly expanded e-Hub contents, architectures, functions and services, development techniques, and application environments. The result is that the definitions of e-Hubs vary significantly due to different services e-Hubs provide and people's views of e-Hubs. Therefore, though under the same umbrella, e-Hubs can be quite different things to different researchers, from a simple B2C (Business-to-Consumer) e-marketplace to a comprehensive virtual enterprise facilitator. For example:

- Cyber Business Centre (URL3) defines an e-Hub as an alternative name for an e-marketplace, and in particular for any sub-category thereof such as a forward aggregator or a reverse aggregator. e-Marketplace is a Business to Business (B2B) online trading forum, often dedicated to

e-business between companies and their customers and suppliers in a particular industry or sector thereof.
- IDS (URL4) describes an e-Hub as a telephone company which passes information from one place to another. The company providing services is invisible to the dealers. e-Hubs can handle the transfer event in either XML- or EDI-based information. Since an e-Hub is an utility in nature, it actually adds nothing to the transaction. As the programs are written to convert data from one place to the next, its work is simply to pass on data.
- Kaplan and Sawhney (1999) define e-Hubs as neutral Internet-based intermediaries that focus on specific industry verticals or specific business processes, host electronic marketplaces, and use various market-making mechanisms to mediate any-to-any transactions among businesses. They create value by aggregating buyers and sellers, creating marketplace liquidity, and reducing transaction costs.
- Mejia and Molina (2001) specify an e-Hub as a business entity, which is responsible for searching opportunities in the global environment and enables the creation of virtual enterprises. The Hub performs the processes of partner search and selection, and configures suitable infrastructures for virtual enterprise formation/commitment (physical, legal, social/cultural, information). To achieve its goals the Hub uses the services provided by virtual industry clusters.

These definitions indicate the different levels of e-Hubs' services. Nevertheless, it can be seen that the core of e-Hubs is that they are Internet-enabled entities which allow users to exchange information for the purpose of value adding. Table 8.1 presents the taxonomy of e-Hubs based on a number of major e-Hub characteristics (Ren et al., 2002). Such a classification is a necessary supplementary to e-Hubs definition, which provides a high-level vision on e-Hubs.

8.3 e-Hubs' services

Either as pure e-marketplaces or complex business entities, e-Hubs achieve their services through Web Services, which represent a revolution with layered services; it enables a dynamic e-business model, fosters collaboration with layered services, and opens the door to new business opportunities.

8.3.1 Collaborative Web Services

Web Services are configured with new technologies such as SOAP (Simple Object Access Protocol), WSDL (Servers, and the Web Services Description Language), WSIL (Web Service Inspection Language), and UDDI (Universal Description, Discovery, Integration). These technologies

Table 8.1 Taxonomy of e-Hubs

Criteria	Classification	Examples
Services provided	Business Hub	e-Hub (URL5), SESAMi (URL6), Covisint (URL7), Agritani (URL8)
	Collaboration Hub	Covisint (URL7)
	Engineering Hub	Engineering e-Hub (URL10)
Purchase situation	Maintenance, repair and operating (MRO) Hub	Ariba, W.W.Grainer, MRO.com, NetBuy.com, BizBuyer.com
	Yield Manager	Utilities, eLance, CapacityWeb.com, and Adauction.com
	Exchanges	e-Steel, PaperExchanges.com, IMX Exchange, and Altra Energy
	Catalogue Hub	PlasticsNet.com, eChemicals.com, Chemdex, and SciQuest.com, and ElectricalWeb.com
Market-making mechanism	Aggregation	Ariba, W.W.Grainer, MRO.com, NetBuy.com, BizBuyer.com, PlasticsNet.com, eChemicals.com, Chemdex, and SciQuest.com, and ElectricalWeb.com
	Matching	Utilities, eLance, CapacityWeb.com, Adauction.com, e-steel, PaperExchanges.com, IMX Exchange, and Altra Energy
Learning ability	Reactive Hub	e-Hub (URL5), SESAMi (URL6), Covisint (URL7), Agritani (URL8)
	Learning Hub	MaldivesRDS.com (URL9)
Host	Internal Hub	www.myaircraft.com
	Coalition Hub	www.covisint.com
	Commercial Hub	www.coprocure.com
		http://www.cablenet.com/index.jsp
		http://www.get-hitech.com/ehub.asp
Attitude	Neutral Hub	e-Hub (URL5), SESAMi (URL6), Covisint (URL7), Agritani (URL8)
	Forward	Ingram Micro, eChemicals
	Reverse Aggregators	FreeMarkjets.com, FOB.com
Expertise	General Hub	e-Hub (URL5), SESAMi (URL6)
	Specialist Hub	eLance, PaperExchanges.com

consist of a model for exchanging XML information, a language for describing services and workflow between business partners, and a directory for finding new business partners, respectively. Together, they enable Web Services for various e-Hubs.

The innovative use of Web Services provides industries with an effective approach to reconfiguring the key business and engineering processes to achieve breakthrough improvements in business opportunities, cost, time, quality, and customer satisfaction. Essentially, they strengthen the ability to communicate, track information and therefore allow different collaborating partners to work on a common set of issues. Table 8.2 lists some of the engineering Web Services.

Table 8.2 Examples of Web Services

Engineering Web Services	Tools available	Possible services supported
Web-facilitated general communication	• Email • Intranet and Internet • Video conferencing • 3D conferencing	Collaboration
Web-facilitated data/information exchange	• Design review • Visualizing • Viewers • Portals • Secure	Knowledge/information management
Web-facilitated coordination	• Workflow modelling • Collaborative portals • Online language translation • Expertise sharing • Group decision-making • Logistics support • ASP/ISP/MSP solutions	Coordination
Web-facilitated data storage	• e-Audits • Information tracing • Data searching • Online database	Knowledge/information management
Web-facilitated legal resolutions	• Authentication • Contract negotiation	Functional
Web-facilitated knowledge management	• Online knowledge retrieval • Online knowledge capture • Online rule bases	Knowledge/information management

8.3.2 *General services offered by e-Hubs*

The services provided by e-Hubs have considerably expanded over the past years. Early efforts focused on enabling users to buy and sell by converting catalogues to HTML and putting up a simple transaction interface. The next level of e-Hubs evolution via XML creates rich environments for execution and coordination of complex business transactions and interactions. A further development is the adoption of e-Hubs to facilitate the activities in the entire value chain with focus on value added services. Morgan Stanley (2000) summarizes e-Hubs' services as five layers: order matching, one to one marketing, content aggregation, transaction fulfilment, and demand and supply chain collaboration (Figure 8.1). Below three major services: e-buying and selling, e-transaction fulfilment, and e-collaboration are discussed.

e-Buying and selling: Traditional e-Hubs such as FreeMarkets.com, e-Bid.co.uk, and eBay.com mainly aim to expand users' business opportunities through matching and aggregating approaches. Some may also facilitate simple business transaction processes such as the declaration

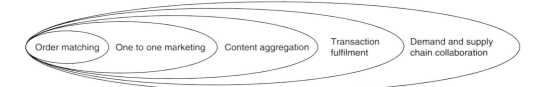

Figure 8.1 e-Hub layers of service

of payment approach and delivery method. These functions, although with limited added value, are still essential for the most comprehensive e-Hubs today. Many of such e-Hubs have expanded their service objects from product to services (e.g. medical or legal advice). Typical of such e-Hub services are:

- Firstly, an e-Hub establishes a relationship between buyer and seller where no relationship existed before.
- Secondly, the e-Hub can provide a mechanism, often via online auctions, by which efficient transactions may be conducted.
- Thirdly, the e-Hub can be an information source, providing useful information about both the products traded and the trades themselves.
- Finally, some e-Hubs may support decision-making by helping participating enterprises identify potential matches and decide if a potential match should be pursued.

e-Transaction fulfilment: With the development of e-Hubs, the customers of e-Hubs are changing from individual (small amount Web shopping, typically B2C) to enterprises (large and complex industrial procurement, typically B2B). The users' requirements have changed considerably and the e-Hubs' services thus have significantly expended. At present, most e-Hubs have moved beyond order matching to supporting the entire transaction fulfilment process. e-Hubs such as Coprocure.com (URL20), GCC (URL11), and PI e-Hub (URL12) can provide comprehensive services to support the overall business transaction process such as identification, evolution, negotiation, and configuration of optimal grouping of trading partners and business processes into a supply chain network so that users can respond to changing market demands with greater efficiency.

e-Collaboration: In most of the B2B marketplaces, relationships are not just about order matching and aggregation, collaboration is particularly important for industrial problems. Industries are pooling significant resources to form e-Hub consortia, where a company can concentrate on its link in the demand and supply chains while taking advantage of recombinant business models. Highly skilled and specialized companies

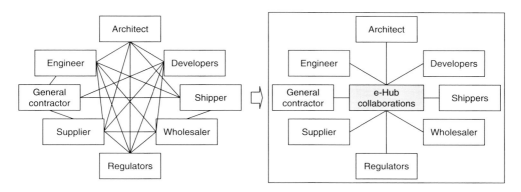

Figure 8.2 From traditional point to point collaborations into e-Hub enabled collaboration in the construction industry. (*Source*: Morgan Stanley, 2000)

collaborate in both synchronized and asynchronized basis as more services can be outsourced within the consortium. e-Hubs (e.g. eHub, URL5; eHub2, URL13) unite businesses and help the traditional point-to-point connections to evolve towards spoke collaborating relationships (Figure 8.2). This improves key processes in the product life cycle such as: improving resource scheduling and allocation; facilitating collaborative product design, creating more efficient inter-enterprise processes; integrating product data within and across enterprise borders and keeping information up-to-date in real time; improving product quality control; and reducing risk throughout the supply chain.

8.3.3 Examples of e-Hubs' services

Table 8.3 lists a number of popular e-Hubs and the major services offered by these e-Hubs. For details, see Ren *et al.* (2002).

A major limitation of the existing e-Hubs is that they are mainly designed to facilitate business processes rather than to support engineering services, particularly in a systematically structured approach. On the other hand, the demands for specialist engineering services are increasing in industries, with growing reliance on outsourced engineering services. Project teams are now characterized as dispersed groups and individuals from different companies and geographical regions. It is therefore, necessary to develop engineering e-Hubs, which can not only facilitate the business process of collaboration, but also be able to tackle the engineering issues of collaboration. eHub1, for example, is successfully attracting business for SMEs in West Midlands area of United Kingdom by considering engineering expertise of each individual SME. eHub1, however, is basically still focusing on partner finding. The engineering e-Hub developed within the EU funded e-Hubs project can provide engineering services that facilitate the project definition and planning between volatile partners.

Table 8.3 Examples of e-Hubs' services

e-Hubs	Basic services provided
e-Hub (URL5)	The largest e-Hub company in the United Kingdom and Ireland, it has developed a series of 20 B2B hubs for the management of supply chains, customer relations, projects, and professional services. • e-Hub procurement • e-Hub tendering • e-Hub quality management • e-Hub project management • e-Hub customer relationship management • e-Hub recruitment
SESAMI (URL6)	A typical business e-Hub, which mainly provides two kinds of services e-procurement and e-auction. It provides the following services in facilitating value chain. • Support: HR, reporting, alliances • Plan: Forecasting, capacity planning • Design: Engineering collaboration, construction, architecture design, design for cost • Source: Vendor selection, supplier qualification • Procure: Purchasing direct and indirect goods, outsource services • Produce: Manufacturing, assembly, construction, forecasts • Sell: Marketing, advertising, sales • Deliver: Shipping, logistics, fulfilment, customization • Service: Post-sales support, customer relations, maintenance
Covisint (URL7)	• Collaboration: Collaboration manager, quote manager • Procurement: Auction, catalogue, asset control • Supply chain: Fulfilment • Quality: Advanced quality manager, problem solver • Portal: Portal, sourcing directory, supplier bulletin, library services
ORACLE (URL14)	• Self-service browser user interface • Workflow • Common data model • Performance measurement • Customizable views • Item cross-referencing • Trading partner cross-referencing • Flexible data exchange format • Flexible user preferences • Communication e-Hub • Bucketed and detailed views on supply and demand • Supply chain collaboration e-Hub • Formalized collaboration partnerships • Constraint detection and exception notification • Collaborative supply planning referencing • Collaborative supply planning • Billing • Global inventory visibility • Vendor managed inventory • Customer managed inventory • Supply chain planning e-Hub • Production sequencing • Exchange marketplace integration • Collaborative demand planning • Multi-lingual support

Agritani (URL8)	• Core business services
	  ○ Procurement
	  ○ Catalogue management
	  ○ Auctions
	  ○ Group buy
	• Value-added service
	  ○ Hosting services
	  ○ Fulfilment/Logistics
	  ○ Financing
	  ○ Catalogue scrubbing
	  ○ Collaboration
	• Marketplace infrastructure services
	  ○ Integration service
	  ○ Marketplace management
	  ○ Messaging

Table 8.3 (continued)

e-Hubs	Basic services provided		
	○ Community/member services ○ Request for quotation ○ Requisition and approval ○ Advertising	○ Reporting and analysis ○ Insurance ○ Inspection ○ Risk management ○ Market intelligence and Industry content ○ Reputation Management ○ Tax and customs	○ Security ○ Workflow
COMPUSOL (URL15)	● RFQ management ● Offer management ● Negotiation ● Agreement templates	● Workflow ● Security ● Notifications/alerts ● Change management	● Audit trail ● Archives ● Net cost analysis ● Service providers
NCMS Private e-Hub (URL16)	● Collaborate on key processes such as design, procurement and production; ● Synchronize market demand with supply chain activities, enabling collaborative demand forecasting such as source, order creation, order routing, order fulfilment, order delivery; ● Transact commerce in real-time – buy, sell, pay settle.		
e-sokoni (URL17)	Provides a facility for businesses to order their non-production supplies: ● Web marketing ● Business process reengineering ● Supply chain management consultancy ● Supplier partnership services ● Inventory management advisory services		
PI e-Hub (URL12)	● Access Service Business Dial Up Account; Small Office Connectivity (ISDN); Corporate Leased Line Access; Broadband SDSL Access. ● Web Hosting DNS Hosting; Biz Web; Internet Data Centre; Managed Services; Outsourced email; Managed Firewall.		
OneChem (URL 18)	● Customer care ● Order management ● e-Hub connectivity ● Contract management ● ERP integration ● Supply chain management		
e-dn.com (URL19)	● Membership management service, ● Data format translation service, ● Business protocol translation service and ● Business flow management service. ● Basic framework for integrating external services with e-Marketplaces.		
Coprocure.com (URL20)	● Search the online product guide featuring over 100,000 IT products ● Save price quotes for future orders or export them for budgetary purposes ● Request products online using the automated purchase requisition process ● Add customers' preferred equipment to virtual Warehouse to standardize IT purchasing company wide ● Utilize advanced administrative reports to easily track assets and purchase history ● Track customers' order status online from placement through delivery		

8.4 Engineering e-Hub

This section presents an engineering e-Hub which has been particularly developed for engineering service outsourcing. The engineering e-Hubs are envisaged to provide an Internet-enabled interface for enterprise-wide product development processes and remote Engineering Service Providers (ESPs). Unlike those e-Hubs collaborating virtual enterprises where e-Hubs play a project manager's role, the engineering e-Hub works much like a mediator which mainly focuses on the activities at strategic and tactical levels rather than at operational level. ESPs undertake the real engineering activities whilst e-Hub facilitates collaboration at different levels and stages.

8.4.1 The role of the engineering e-Hub in project preparation process

Good project preparation and planning is vital for the effectiveness of dispersed collaborative engineering teams, thus adding value in current collaborative engineering platforms. It is this realization that has companies looking for support to initiate and plan partnerships that are remote, time-critical, and volatile. Such partnerships necessitate a new generation of collaborative project preparation and planning methodologies and services. Unlike the existing business, engineering and collaborative e-Hubs (e.g. e-Hub, URL5; SESAMI, URL6; ORACLE, URL14; COMPUSOL, URL15), the e-engineering Hub (URL10) was developed to offer collaborative project planning (PP) services focusing on the collaborative, tactical decision-making that goes into the formation, work planning, contracting, and trust building.

As illustrated in Figure 8.3, the e-Hub offers a shared workspace for consortium members to conduct collaborative PP; thus it is possible to integrate all their information, expertise and planning efforts together to improve the efficiency and effectiveness of PP and reduce the risks caused by lack of information and wrong decision-making. The collaborative work plan defined through such an integrated PP process provides a firm foundation for the project contract negotiation between client and ESPs and further project execution.

The specifically targeted end-users for the e-Hub are project managers. Unlike the top management level or the operational level in an organization, project managers are the embodiment of a variety of tactical functions that are sandwiched between their masters at strategic levels and the minions whose development is to be planned and guided. A major function of the tactical level is to shape the operational level in such a way that the objectives of the strategic level are fulfilled. Another major function is to ensure that the strategic level has access to the information it requires to support its own decision-making and reporting processes. Thus, the tactical level is more or less a connector between the operational

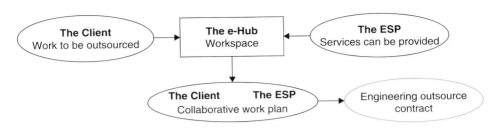

Figure 8.3 The collaborative/intermediating role of e-Hub

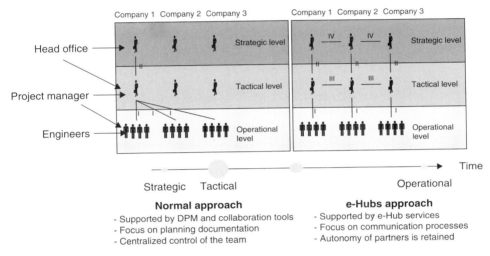

Figure 8.4 A middle-up-down view of project management. (*Source*: Augenbroe and Crehan, 2004)

and strategic level. The e-Hub is designed to facilitate the collaboration at the tactical management level (Figure 8.4).

There are three main constraints in the e-Hubs study:

(1) The e-Hub works in the scope of selected 'project windows' (i.e. the work to be outsourced) rather than an overall project.
(2) The e-Hub lifecycle includes five stages: (1) deciding what to partner; (2) finding the right partner; (3) agreeing with that partner what, how, when to do it; (4) executing the plan developed; and (5) dissolving the partnership. This study only focuses on stage 3.
(3) This study is mainly concerned with the automatic or semi-automatic services of the e-Hub rather than the role of the e-Hub staff or their interactions with other consortium members.

8.4.2 e-Hub's functional architecture

The development of the e-Hub is backed by multi-level knowledge and technologies such as collaboration platforms offering shared project workspaces for team building, group communication, project management methods, portal 'store front' functionality for marketing, contract management, process representation, sharing and execution, knowledge capturing and sharing. These technologies, supported by available engineering Web Services, form the basis of the e-Hub. With a properly designed functional architecture, they can be integrated seamlessly to form a collaboration gateway that provides a systematic, transparent, traceable and controllable, and effective collaboration approach.

Theoretical basis of collaborative project planning

PP is paramount for successful project management; the basic methodology for PP is well known, proven by years of experience, and supported by many well-developed tools (Ren *et al.*, 2006). Some of the important characteristics of PP can be summarized as (PMI, 2000): PP is structured and predictable; there are generic steps/parts of PP; also, the logic of PP is deterministic. PP is comparable in each particular project type. A project planning model (PPM) was developed based on these characteristics, and forms the basis of the e-Hub's functional architecture. Three fundamental issues have to be addressed when studying collaborative PP:

(1) *The content of a project plan:* The objective of classic PP is to define project scope, timeframe, budget, and other key issues of project. With visible targets and constraints, PP defines a set of formal subtasks providing optimal resource allocation, and control and management of issues related to various aspects of a project. As a result, a number of sub-work plans are generated at different stages, such as project summary, project charter and general scope statements, work breakdown structure, schedule, estimated cost, resource plan, delivery plan, risk plan, and quality plan. As most of these plans have predictable and generic components, various templates (e.g. URL10, URL11, URL12) have been developed to summarize the contents of each plan. Also, standards institutions, agencies, and large corporations often have their own well-defined formal document templates to address these plans. By adopting these templates, users – even those not experts in PP – are able to address the key issues of a project plan.

(2) *Process of project planning:* PP normally has two main phases – preliminary planning (process to generate expression of interest, draft of a business plan, analysis of technical feasibility, or potential deliverables) and detailed planning (in-depth study to create plans for process quality, finance management, quality assurance, and time-line scheduling). The Project Management Body of Knowledge (PMI, 2000) classifies PP as *core processes* (e.g. scope planning, activity definition, schedule development, resource planning, and cost estimating) *and facilitating processes* (e.g. quality planning, communication planning, and risk planning). The core PP processes have clear inter-dependencies and

are thus generally performed in the same order in the majority of projects, while the facilitating processes are dependent on the nature and structure of the project. Such general PP processes are considered as generic and structured. Based on these theoretical studies, a generic PPM for engineering service outsource projects has been developed. By following the PPM, and using related attribute templates, SMEs are able to define and negotiate the details of the engineering services.

(3) *PP dedicated collaboration:* PP offers an opportunity for project participants to share and balance their objectives, resources, expertise, and constraints. PP generally has a form of iterative loop, or of a dialogue, in which client's requirements and provider's proposals are continuously discussed and gradually refined. Clients, usually have problems with clarification of what is possible and what is desirable to expect from the project in return for invested resources, while providers want to balance available resources and expected efforts, associated with fulfilment of a client's requirements. The generic PPM provides structured guidelines for SMEs to collaborate, detailing what should be defined at which stage.

Supporting technology

Best-of-breed technologies for Internet-based communication, team collaboration, and operational e-engineering form the baseline of the e-Hub development. They are positioned at the core of the e-Hubs. The specific information technologies deployed for the PP extensions are workflow management system and a basic collaboration platform.

Workflow management system for process management

WfMC (2000, 2001) defines workflow as: 'The automation of a business process, in whole or part, during which documents, information or tasks are passed from one participant to another for action, according to a set of procedural rules'. Gerogakopoulos *et al.* (1995) give a more explicit description of a workflow process. They define a workflow as a collection of activities organized to accomplish some business goals. An activity can be performed by software system(s), human(s), or a combination of these. In addition to a collection of activities, a workflow may include constraints that influence the order of performing activities as well as information flow between them.

The essential workflow characteristics are persons, activities, application tools, and resources. Marshak (1994, 1997) defines the '3Rs' and the '3Ps' of workflow technology:

Rules: Workflow systems take various business rules into account.

Routes: A route is strongly coupled to the concept of information logistic that typically supports organization flows of all kinds of objects including documents, forms, and processes.

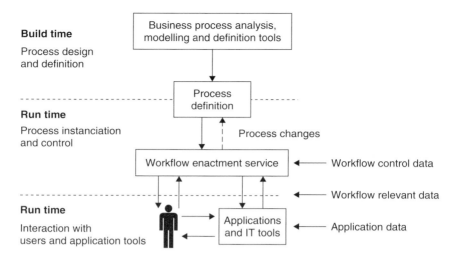

Figure 8.5 **Workflow system architecture and data structure.** (*Source*: WfMC, 2000, 2002)

Roles: Information is routed to roles rather than to a particular person. The role in an organization is a group of people with the required skills and authority.

Processes: Business/engineering processes span over organization units and legacy information systems.

Policies: Policies correspond to a normative process model that describes how certain processes should be handled.

Practices: This is the way that work is actually performed in the organization.

A Workflow Management System (WfMS) supports the specification, execution, and dynamic control of workflows involving humans and information systems (McCarthy and Sarin, 1993). Figure 8.5 illustrates an outline of a WfMS architecture. The process analysis, modelling and definition tools facilitate the specification of the components of a workflow as a process definition. The workflow enactment service enacts a process definition by assigning tasks to humans and software systems while also maintaining constraints between tasks. The workflow control data represents the dynamic state of the workflow system and its process instance, which is managed and accessible by the workflow management system. The workflow relevant data is used by the WfMS to determine the state transitions of the workflow instance.

The three major solutions which a WfMS provides to business and engineering problems are summarized as follows (Prior, 2003):

- A unique and systematic approach to model business or engineering processes were the key features involved in the business or engineering

processes such as roles, activities, inter-dependencies, routes, and resources are considered.
- An effective tool to administrate contract management.
- An approach to facilitating knowledge management, records management, and process monitoring.

Technologies for collaboration

To facilitate collaboration, a Basic Collaboration Platform (BCP) has been developed in the e-Hubs project, built on Java and J2EE technology. The BCP is a best-of-breed collaborative virtual environment that provides two levels of services: common services and engineering services. Some of the key features of the common services include:

- *User Management*: registration, building user profile, access controls via groups/roles, directory.
- *Collaboration Features*: email, forum, online chatting, newsgroups, discussion groups, calendar, and meeting support.
- *Document management*: upload and download, access right, version control, copy right, and Enterprise Applications Integration (EAI).
- *Internationalization*: portal support for different character sets, portal support for different date formats and time zones.
- *Security*: encryption of key data as required, access control, restricting access to data and documents, cross-authentication or single-sign-on mechanisms.
- *Others*: browser compatibility, Graphical User Interface (GUI), searching, access to Web Services, mobile/wireless computing, openness.

In the BCP a set of activity components are added to support specific needs. Among them are a PP whiteboard, annotation function, and a specific contract editor as example of embedded external services.

- *The PP Whiteboard* allows users discussing a particular section of project plan online both textually and graphically. Users can import screenshots from planning documents or use a drawing toolbox to create sketches, use online chatting and drawing to discuss the details, and save the animated discussions into files.
- *The annotation function* allows users marking and explaining changes on existing standard documents. This is helpful to make the e-Hub a transparent and traceable environment for joint work plan definition and negotiation.
- The BCP is also able to support and integrate some *external services* such as eLEGAL (URL21) contract editor and Ganttproject (URL22). Although such services are not technically integrated as an e-Hub service, the logical binding between the e-Hub and contract editor is achieved through the use of customizable workflows.

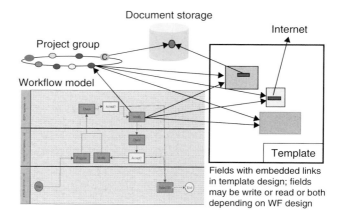

Figure 8.6 Functional architecture of PPM

Functional architecture

The functional architecture of the e-Hub, which harmonizes the group collaboration functionality provided by the BCP with the dedicated PP functional components, is described in this section. Collaborative PP is viewed by the e-Hub as a managed collaborative process that transparently generates a comprehensive 'project plan', consisting of a set of interlinked documents. Documents are either structured models or unstructured (text) information. The added value of the e-Hub is that the generation process is composed of logically ordered activities that drive the collaborative generation of (partly) structured content. The logical ordering of activities and their relationship to structured content is embodied in a formal PPM as explained above. The PPM is a set of WF models, each of which operates on one or more content templates. Figure 8.6 shows the functional architecture for this.

Figure 8.7 shows a workflow model that defines that a task is to be assigned to a particular project planner (a member of the project group), while three fields of the information template have been defined to be accessible by this task. At the time of enactment, the project planner that is assigned this task will fill these fields with specific information. In doing so, s/he may use links to internal documents and external resources whenever appropriate, as indicated in Figure 8.6. It is very important that workflows can be started at will and run concurrently, reflecting the multi-tasking and multi-threading working styles of real-life PP. Planners typically work on different issues concurrently (Figure 8.7).

Figure 8.7 shows the typical situation where a PP team deploys a PPM with four workflows. In this case three workflows have executed, or are executing concurrently (WF1, WF2, WF3). WF3 has been uploaded in the workspace but has not been executed, whereas WF2 is executing but has

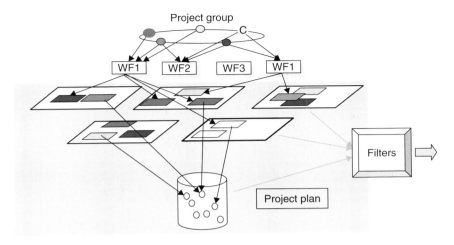

Figure 8.7 **Workflows operating on a collection of templates**

not yet reached the state where interaction with information templates has taken place.

In the figure, it is assumed that the four workflows have been defined to interact with five different information templates. Each template has a set of fields and different members of the project team have inserted information in these fields as mandated by the logic in the workflows (the coloured fields). It is also indicated that fields may contain hyperlinks to stored documents and other fields thus making the space of project documents (the project plan) a rich information resource, which can be navigated in many different ways. It is now possible to define filters in the e-Hub workspace that transform the generated project information into structured reports, contract templates, and project execution schedules.

Figure 8.8 shows the high-level architecture of the e-Hub distinguishing the five major functional modules and how they operate on communities (containing client, providers, external consultants, etc.), document, workflow logic, and templates.

The above figure shows how the community manager module (managed by an administrator) maintains the identities of all users, and assigns access rights to engineering domains and any active project within the engineering domains (managed by a domain coordinator). Every engineering domain has an open marketplace, to which every registered domain user has automatic access.

The communication manager module controls chat, email, Web meeting, and other events; whereas the document manager provides access to document. The workflow management module stores and manages the PPM, creates instances of workflows, and exports instances to the WF run-time environment. Coordinators of project groups decide which PPM, or which packages within a PPM will be made available to the project

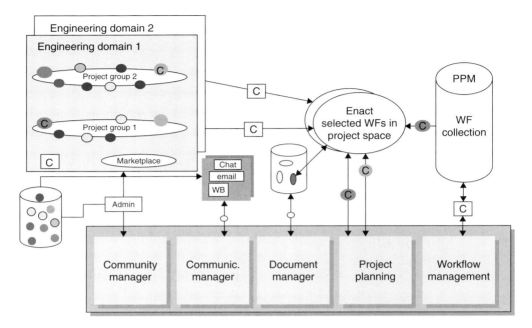

Figure 8.8 Functional architecture of the e-Hub

team. The PP module enacts an instance of a workflow, allows the assignment of actual project partners to roles defined in the WF, maintains the link between WF variables and native e-Hub applications, and manages the link between WFs and information templates.

The above architecture combines the functions for community control, process control, and managed generation of information in the predefined templates. It is this mix that enables a structured PP dialogue, that is arrive at a joint resolution among partners about tasks, responsibilities, dependencies, incentives, penalties during project design, etc.

It is one of the longer-term objectives of PP communities to harness the commonality of PP steps and store them as PPM templates. The e-Hub business entity could act as the custodian of these templates and make them available to new PP teams in the same domain. If the PPM templates contain proprietary client knowledge it is more likely that the client will retain the intellectual property and will make models only available to internal project teams. In either case, the e-Hub business entity will offer services (either public or for specific clients) that allow the cultivation, extension, and testing of PPM templates.

8.5 Engineering services

Based on the functional architecture, the e-Hub provides a series of engineering services, which have been tested in both construction and manufacturing engineering test beds. In the construction scenario, a Dutch design firm needs to subcontract the seismic risk analysis work to a seismic analysis consultant located in Southern Italy through the e-Hub. In this case, the Client and the ESP need to define three major issues:

- *Cost estimate*: Both parties are concerned about the project cost. If they cannot reach an agreement, they would not carry on the negotiation. Before they can negotiate the project cost, however, they need to reach an agreement on some of the initial project requirements.
- *Project schedule*: The Client needs to address a number of detailed execution plans including: schedule, quality plan, and change protocol. Of these, the Client is particularly interested in project schedule because s/he needs to arrange other related activities accordingly.
- *Contract and conditions of contract*: The Client and the ESP need to finalize the contract and related conditions.

The Client and the ESP conduct the collaborative PP on the e-Hub platform. The e-Hub provides a workflow and document template for each item. Due to the space limitation, this chapter will only demonstrate the workflow and attributes template related to cost estimate. Figure 8.9 illustrates the meta workflow and attribute templates for project description and cost estimate.

After defining all the execution plans, the Client and the ESP enter the contract negotiation stage, which includes two stages: negotiation of agreement and negotiation of conditions of the contract. A standard engineering service outsourcing agreement template is adopted. A contract negotiation workflow is developed to facilitate the negotiation of the key contract items in the agreement template. In this case, there are four key items (i.e. number of test samples, final service cost, liquidated damage, and governing laws) to be addressed in the agreement template, the contract negotiation workflow guides the Client and the ESP to negotiate these items step by step (e.g. number of testing sample, service cost, liquidated damage, governing laws). The items finally agreed by both parties will fill into the 'right place' in agreement template in the database (Figure 8.10). There are two particular advantages of this approach:

(1) The agreement template highlights the key contract items based on different engineering services; the enactment of the contract negotiation workflow thus guide users to negotiate these key issues.
(2) The collaborative work statements generated through previous workflows are integrated into the agreement template, which provide a sound basis for the service outsource contract.

Besides the above core engineering service, the e-Hub also offers a number of engineering services at different levels. For example, the e-Hub allows

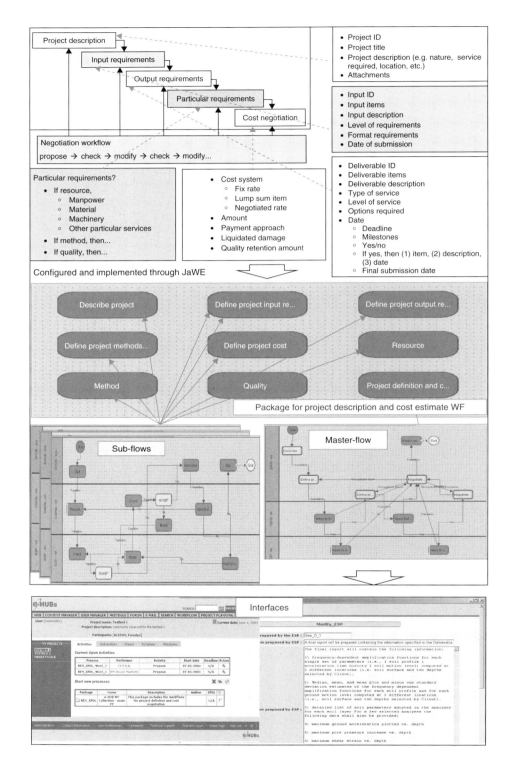

Figure 8.9 Project description and cost estimate workflow

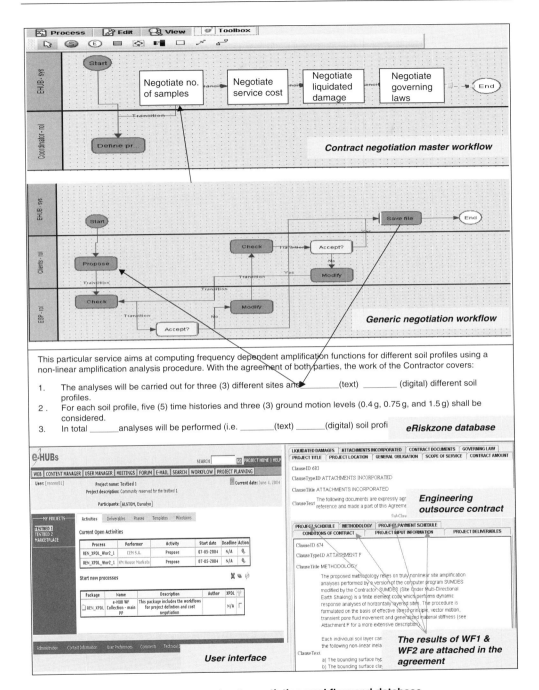

Figure 8.10 Interaction between contract negotiation workflow and database

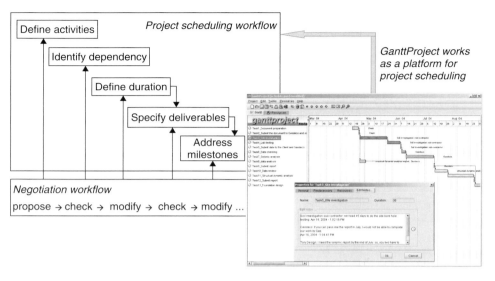

Figure 8.11 Scheduling service incorporated with the e-Hub engineering services

users to adopt best breed engineering tools to conduct collaborative PP, for example GanttProject is adopted in project scheduling workflow (Figure 8.11) and the eLEGAL contract editor is adopted as a negotiation platform for project participants to address the complex conditions of contract (for details, see Ren *et al.*, 2004, 2006).

8.6 Problems and challenges

Although e-Hubs have created great benefits to businesses and industries, many problems have been encountered and need to be solved in the development and application of e-Hubs. Customers have taken full advantage of e-Hubs. According to Stevens (2001), the empirical results show that:

- B2B e-Hubs are unlikely to enable a reduction in overall transaction costs, sufficient to facilitate entry into new global markets.
- B2B e-Hub providers do not appear to be providing the types of services that firms need to engage in transaction preparation and completion.
- Most e-Hubs focus on minimizing information search costs – bringing 'stranger' together – 'buyer and seller beware'.
- Product certification, product quality, or trading partner reputation is not easily accessible at e-Hubs.

The reasons for these are multi-fold. Essentially, several aspects are lacking in terms of using Web Services for e-souring and enterprise integration such as:

- Unclear e-Hubs development strategy.
- Limited or no control over the integrated business logic that involves other partners or enterprises.
- No service provision and subscription capability.
- Lack of efficient UDDI search for e-business integration.
- Lack of relationship defined in current Web Services specification for integration.
- Lack of message tracking.
- No inherent scalability.
- No inherent security.
- Lack of effective transaction support.
- Unclear trust and legal aspects.

8.7 Conclusions

Business e-Hubs are growing very rapidly due to the great benefits they offer to users and developers (i.e. the ability to buy, sell, advertise goods and services, and collaborate value chains through e-Hubs). To obtain a high-level view of e-Hubs, this chapter reviewed several key aspects of e-Hubs including definition, classification, services and problems, and potential challenges with emphases on e-Hub's services.

e-Hubs, as a general platform, greatly enhance business opportunities and improve business and engineering process. This global network of servers is used by many industry leaders to liquidate their service products and establish exchange networks that they now call e-Hubs. e-Hubs have clearly demonstrated the power of the Internet to perform as a real-time, global distribution network by dissolving the constraints of time and geography and made it possible for buyers, business-to-business, business-to-consumers, and business-to-industry to become more fully engaged. e-Hubs create value by aggregating buyers and sellers, creating marketplace liquidity, and reducing transaction costs and time. Essentially, e-Hubs' services can be regarded as three levels core services and value-added services.

Core services: Most e-Hubs can provide services to increase business opportunities which are the core services e-Hubs can offer. Such services include: procurement, marketing, quotation requests, offer management, tender, auctions, catalogues, asset control, community/member services, requisition and approval, and advertising.

Value-added services: Besides the above core services, most of today's e-Hubs also provide some other services which can add much value to the entire value chains. These value added services can be classified as two layers.

- The first layer services mainly facilitate the entire business transaction process such as: messaging, hosting services, fulfilment/logistics, billing and settlement, financing, catalogue scrubbing, agreement templates, customer care, security, insurance, and tax and customs.
- The second layer services enable e-Hubs to integrate virtual clusters (mainly SMEs) into virtual enterprises by collaborating the information among the virtual clusters both vertically and horizontally. Great value can be generated through the virtual enterprises. Such services typically include: itegration service (e-Hub community, external entities, other e-marketplaces), collaboration (demand, supply), public discussion/forum, reporting and analysis (historical transaction information of all products, financial reports, trends and patterns for demand forecasting, order fulfilment rates), workflow management, quality control, business process reengineering, contract management, supply chain management, problem solvers, library services, and market intelligence and industry content.

The engineering e-Hub is a novel concept in the e-Hub domain. By adopting a systematically developed functional architecture, the e-Hub is able to:

- Offer job procurement through the seamless connection between the engineering portal and the e-Hub, contracting and collaborative process facilities, including handshaking, process sharing, and process mediation.
- Offer configurable e-engineering process templates, thus harnessing proven procedures for remote collaboration.
- Enforce quality assurance in collaborative project preparation and planning through transparent procedures and standard practices, thus paving the way for the improvement of quality assurance in the project execution process and, hence, for the overall collaborative engineering process.
- Enhance trust building for SMEs to adopt e-business and e-engineering practice. The e-Hub provides a unique trust building approach that is to build trust by offering SMEs the transparent collaborative PP platform. This is particularly meaningful as most of the current e-business trust building studies only focus on the security perspectives while neglecting engineering issues themselves.
- Offer low entry barrier for SMEs to the global marketplace for the outsourcing and fulfilment of engineering sub-tasks.

Table 8.4 lists a few major areas for the development of e-Hubs.

Table 8.4 Further studies of e-Hubs

Field	Tradition	Future
Functionality	Match-making or transaction execution	Collaborative end-to end value chain likens such as design, sourcing, and sales/marketing
Types of interactions	One-off, arms-length	Long-term, partnership-focused, value chain integration
Participants	Competitors in consortia	Supply chain partners
Source of value	Market efficiency Process standardization automation	• Collaborative process improvements impacting • Non-price levers of total cost of ownership • Costs in multiple aspects of value chain
	Responsive	Intelligent
	Static	Dynamic

Acknowledgements

The e-Hubs project was supported by the European Commission under the IST programme (Contract no: IST-2001-34031). The authors would like to acknowledge the financial support of the European Commission, and record their appreciation to the e-Hubs project partners for their contributions to the project.

References

Gerogakopoulos, D., Hornick, M. and Shet, A. (1995) An overview of workflow management: Form process modelling to workflow automation infrastructure. *Distributed and Parallel Databases*, 3(2), 119–153.

Kaplan, S. and Sawhney, M. (1999) Available at http://www.umsl.edu/~viehland/IS491week6.html.

Marshak, R.T. (1994) Workflow White Paper: An overview of workflow software. In: *WORKFLOW'94 Conference Proceedings*, San Jose, USA.

Marshak, R.T. (1997). InConcert workflow: Independent from XSOFT, InConcert Inc. provides flexible workflow underlying engineering team support. *Workgroup Computing Report*, Patricia Seybold Group.

McCarthy, D.R. and Sarin, S.K. (1993) Workflow and transaction in InConcert. *Bulletin of the Technical Committee on Data Engineering*, Special Issues on workflow extended transaction systems, IEEE **16**(2), June.

Mejia, R. and Molina, A. (2001) Virtual enterprise broker: Processes, methods and tools. *Progress Report*, CSIM-ITESM.

Morgan Stanley Dean Witter Internet Research (2000) Collaborations in action, hubs and spoke always beats point-to-point. *Collaborative Commerce*, April.

Prior C. (2003) Workflow and process management. In *Workflow Handbook 2003* (L. Fischer ed). Future Strategies Inc., Florida, pp. 17–25.

PMI (2000) *A Guide to the Project Management Body of Knowledge*. Project Management Institute Inc., Newtown Square, Pennsylvania, USA.

Ren, Z., Anumba, C.J. and Hassan, T.M. (2002) A review of e-Hubs and their services. *Technical Report*, Loughborough University, UK.

Ren, Z., Hassan, T.M., Anumba, C.J., Augenbroe, G. and Mangini, M. (2004) Electronic contracting in the e-engineering hub. In: *The Proceedings of the 5th IFIP Working Conference on Virtual Enterprises (PRO-VE 2004)*, Toulouse, France, pp. 235–244.

Ren, Z., Anumba, C.J., Hassan, T.M., Augenbroe, G. and Mangini, M. (2006) Collaborative project planning: A novel approach through an e-engineering hub – A case study of seismic risk analysis. *Computer in Industries*, **57**(3), 218–230.

URL1: http://www.itnetplc.com/e3/internet/eb/news.nsf/MasterNewsFrameset!OpenFrameSet&Frame=body&Src=_u5tfluc1g68qjce 1o6oo30d1jc 4qm8p9edppmcbpg5th66dpmcoom6e1k6osj2o9m68o30c1i6kr3ge9 l60o38oj2cks368afe1imsh3fcdqmqpbeegj4gqb7d1m6ipr8eguj4b355 lk7aoh685qn8rq6e9gmqpb4_ (last access: Dec., 2005).

URL2: http://www.gca.org/attend/2000_conferences/ebImplement_2000/casestudies_thursday.htm.

URL3: http://www.nottingham.ac.uk/cyber/G116.html.

URL4: http://www.idsastra.com/about/articles/ebusiness.htm.

URL5: http://www.e-hub.com/pages/home.asp.

URL6: http://www.sesami.com/index1.html.

URL7: http://www.covisint.com.

URL8: http://www.agritani-hub.com.

URL9: http://www.maldivesrds.com/resorts/main.html.

URL10: www. e-Hub.org.

URL11: http://www.cablenet.com/GCC.

URL12: http //www.pacfusion.com/corporate/services-ehub.shtml.

URL13: http://www.nb2bc.co.uk/home/.

URL14: http: //www.estafeta.com/ingles/comercioe/b2b.html.

URL15: http: //www.compusolsoftware.com/ehub.htm.

URL16: http: //www.techcon.ncms.org/00fall/presentations/BtoB-Haynes.pdf

URL17: http: //www.e-sokoni.com/products.html.

URL18: http://www.onechem.com/.

URL19: http: //www.kawasaki.com/kengine/press/news11.html.

URL20: http: //www.coprocure.com.

URL21: http://cic.vtt.fi/projects/elegal/public.html.

URL22: http://ganttproject.sourceforge.net/.

Stevens, D. (2001) *Boom, Bust, and e-Hubs*. Available at www.backbonemag.com.

WfMC (2000) Proposal for an asynchronous HTTP binding of WF-XML. *Workflow Management Coalition*. Future Strategies Inc.

WfMC (2001) The WfMC glossary. *Workflow Handbook 2001*, Future Strategies Inc.

WfMC (2002) The workflow management coalition workflow standard, workflow process definition interface – XML process definition language. *Workflow Management Coalition*, Future Strategies Inc.

9 Web Services and aecXML-Based e-Business System for Construction Products Procurement

Stephen C.W. Kong, Heng Li and Chimay J. Anumba

9.1 Introduction

The advancement of related Web technology and services has provided a rich environment for developing many e-business systems for construction products procurement to improve business processes, to cut administration costs and provide more comprehensive information. Currently, these e-business systems are owned and operated by many different parties such as manufacturers, suppliers, agents and application services providers. These systems are non-interoperable which creates problems and difficulties for the buyers who use these systems to purchase construction materials. Moreover, the great potential of using the Web as a global repository to provide rich information is lost in this non-interoperable structure. In this chapter, the concept of Electronic Union (E-Union) is presented, which integrates the information and services provided by different e-business systems for construction products procurement. The underlying technologies such as Web Services and XML for implementing E-Union are then introduced. The last part of the chapter describes the Web Services and aecXML-based framework of E-Union and a prototypical implementation.

9.2 The need for e-procurement of construction products

Construction products typically account for 40–45% of the cost of all construction work (Agapiou *et al.*, 1998). It is important to ensure that construction products are procured at the right price, quality and time. On the other hand, architects require comprehensive products information for their building design works. Construction products information is needed throughout the whole life cycle of a building, including the design, construction, maintenance and demolition stages. Provision of comprehensive and updated construction products information to construction industry practitioners is vital to the success of building projects. Over the last few years, the Internet has evolved from being a scientific network only, to a platform that is enabling a new generation of business. More and more companies and organizations are doing different types of business and offer value-added services on the Internet (Hoffman *et al.*, 1995). While the first stage was fuelled by the vision and innovation of business-to-consumer (B2C) Internet companies, the current phase is defined by the leadership and market success of companies engaged in business-to-business (B2B) e-business (Sculley and William, 2000). The B2B e-trading marketplaces, which allow large communities of buyers and suppliers to meet and trade with each other, are an essential component of B2B e-business applications. They resemble stock exchanges in many ways, including the way they are set up and organized and the trading methods they employ – but they are trading physical commodities such as doors, tiles and steel. They enable many-to-many relationships between multiple buyers and sellers in the construction industry, who come together and find each other in cyberspace. They allow participants to access various mechanisms to buy and sell almost anything, from services to direct materials. Buyers and suppliers leverage economies of scale in their trading relationships and access a more 'liquid' marketplace. Sellers find buyers for their goods, and buyers find suppliers with goods to sell.

Many-to-many liquidity allows the use of dynamic pricing models, thus further improving the economic efficiency of the market. Construction companies are now conducting their business using Web-based e-business systems. Many believe that e-business can provide a win-win situation for both suppliers and buyers, as it can provide an expanded marketplace within which buyers and suppliers can communicate directly with each other. e-Business might bring the answer awaited by clients or construction firms, that is, to create the solution for the procurement of construction products using non-traditional methods, avoiding delays, high prices, lack of specified products, etc. (Castro-Lacouture *et al.*, 2001). Online construction trading markets are not limited by the physical limitations of store spaces and can carry a much larger variety of products and different styles and sizes. At the same time, buyers can search through a wide range of products with low transaction costs at any time convenient to them. More importantly, the direct communication between buyers and suppliers will cut off the multiple layers of middlemen between suppliers and buyers. These middlemen take commissions and fees from both buyers and suppliers. The use of e-business will therefore directly benefit the buyers so they can efficiently purchase cheaper products with a variety of choices (Bakos, 1991). The last few years have witnessed the emergence of online B2B e-trading marketplaces. Well-known examples include Catex, Chemdex, e-STEEL, Metal-Site and VHCome (Kong *et al.*, 2001a). Currently, however, it is often the case that, within a particular industry such as the construction industry, many e-trading marketplaces have been developed, owned and/or hosted by different companies. Each of the e-trading marketplaces forms a closed system with their own customers and clients. The totality of these e-trading marketplaces appears to be islands in the sea, as they are isolated and with no interoperation between each other. In this chapter, a framework is presented for developing an interoperable e-trading marketplace for the construction industry by linking all the existing e-trading marketplaces, as any single e-trading marketplace may not be able to meet all the requirements of the buyers. The ability of marketplaces to interoperate extends the idea of liquidity and network effect by joining more buyers with more suppliers without sacrificing the ability of each marketplace to be highly specific to the supply chain node or target buyer group it serves. The concept of E-Union is described, which integrates the services provided by different e-trading marketplaces in the construction industry to provide an open and unified e-trading marketplace, which is enabled by the use of Web Services technologies and information standardization.

Web Services, a kind of distributed systems technology, can allow one machine to access/invoke methods on other machines via common data formats and protocols such as Extensible Markup Language (XML) and Hypertext Transfer Protocol (HTTP). This technology enables sharing of construction products information stored in different computers. By introducing a common standard of construction products information representation such as aecXML, e-business systems providing construction products information can utilize Web Services technologies to communicate with each other, thereby linking separated islands of information.

9.3 Existing e-business systems for construction products procurement

A typical e-business system for construction products procurement consists of an electronic products catalogue and functions for conducting online transactions. Electronic products catalogues shared through the Internet allow customers to browse through multimedia product representations and to get relevant information concerning the product (Timm and Rosewitz, 1998). It is also the reference for product selection that can assist with source selection and description of terms and conditions (Keller and Genesereth, 1997). Stanoevska-Slabeva and Schmid (2000) define Internet-based electronic products catalogues as interactive multimedia interfaces between buyers and sellers on the Internet, which support product representation, search and classification and have interfaces to other market services such as negotiation, ordering and payment. Internet-based electronic product catalogues can also be seen as an interactive front-end interface that provides classified and structured product information, and supports product searching, comparison and evaluation, and may have linkages with other e-business services such as bidding, ordering and payment (Kong *et al.*, 2001b). Web-based electronic products catalogues also utilize the Internet as the network for transmitting product information. In addition, it uses a Web browser as the front-end interface for accessing product information.

9.3.1 Types of construction product information

Product information in electronic catalogues can be classified into two types, structured and unstructured information. Structured information is stored in relational databases and can easily be offered online with available state-of-the-art databases and online merchant technologies (Lincke, 1998). The form of unstructured information varies from documents to complex multimedia data structures (Ellsworth and Ellsworth, 1995). For instance, construction product categories and product attributes (such as price, dimension, brand name, etc.) are usually stored structurally in the database. Product images, CAD drawings and test reports are typical examples of unstructured construction product information. Besides information describing properties of the product itself, other information related to the product such as user feedback, suppliers of that product, pricing and discount, etc. are also available in Web-based product catalogues. These kinds of information facilitate product evaluation and selection, and help buyers make better decisions in their purchasing jobs.

9.3.2 Architecture of e-business system for construction products procurement

e-Business systems for construction products procurement are multi-tier applications, which divide functionality into separate tiers. These tiers can be located on the same computer, but they typically reside on separate computers. Figure 9.1 shows the basic structure of a three-tier Web-based application. The three tiers are client tier, middle tier and information tier.

The information tier maintains data pertaining to the application. This tier typically stores data in a relational database management system such as Microsoft SQL server. Structured construction products information is stored in this tier. Product information or other related information can be stored in multiple databases, which provide the data needed for the electronic products catalogue.

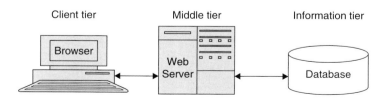

Figure 9.1 Three-tier architecture of Web-based application

The middle tier acts as an intermediary between data in the information tier and the application's client. It implements business logic, controller logic and presentation logic to control interactions between application clients and application data. The controller logic processes client requests, such as requests to view a specific type of construction product in a catalogue, and retrieves data from the database. The presentation logic then processes data from the information tier and presents the content to the client typically in the form of hypertext markup language (HTML) documents. Business logic in the middle tier enforces business rules and ensures that data are reliable before the server application updates the database or presents data to users. Business rules dictate how clients can and cannot access application data and how applications process data. For instance construction product information in the electronic catalogue may only be available to registered members, and members cannot view each other's information. Also, before new construction product information is added to the database, it is necessary to validate the format of this information and to make sure that this information is provided by authorized users.

The client tier is the application's user interface, which is typically a Web browser such as Microsoft Internet Explorer or Firefox. Users of an e-business system interact directly with the application through the user interface, which in standard HTML format is quite limited. However, with the addition of Java technology or other third party plug-ins in the Web browser, there are unlimited ways of building a user interface. The client tier interacts with the middle tier to make requests and to retrieve data from the information tier. It then displays to the user the data retrieved from the middle tier.

9.4 Limitations of existing e-business systems

An e-business system for construction products procurement basically contains two major functions, providing construction product information and facilitating trading transactions. Construction product information, depending on the design of the system, can be provided by product suppliers, agents or manufacturers. Some e-business systems are owned and operated by a manufacturer or a supplier and so their product information is very limited. Systems that are owned by an agent company or an application service provider contain more product information provided by different manufacturers and suppliers. In both cases, the process of purchasing products from these e-business systems is similar. Buyers firstly log into the system and browse through the product catalogue or search for products by specifying search criteria such as brand, model, quality and price. When suitable products are found, buyers can place an order to buy from the suppliers or to select an agent to arrange purchase of all the necessary products for them. Different e-business systems are operated by different types of organizations and they attract different groups of buyers. They specialize

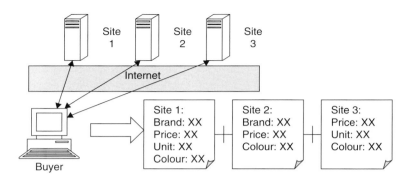

Figure 9.2 Traditional approach to searching and buying a product in e-business websites

in trading products from suppliers in different regions. Usually, a buyer needs to visit more than one of these e-business systems in order to find all the necessary construction products. Figure 9.2 illustrates this situation. Different Websites have different product search and display patterns. They also provide different attributes of construction products. The variety and heterogeneity of different e-business Websites create problems for the buyers. Finding products in these sites requires buyers to acquire and maintain a list of Web addresses, to interpret and understand the semantics and navigation methods in different sites, and to manually integrate product information from these sites for evaluation. This kind of product finding can be time-consuming and the buyer is required to keep abreast of new sites. From the point of view of sellers, a closed e-business system cannot retrieve information from other systems and thus sellers cannot get comprehensive market information for making decisions on production and distribution.

9.5 The E-Union concept

To provide better value-added services to the buyers and sellers, the concept of E-Union has been developed by linking together relevant e-business systems so that communication and information sharing between these systems can be facilitated (Li *et al.*, 2002). In the E-Union framework, different construction products trading sites are joined together by an application provided in their information system for intercommunication and information exchange. As illustrated in Figure 9.3, a buyer using one of the e-business Websites cannot only get product information from this site but also product information stored in other sites. All e-business systems under the framework are linked together by an E-Union server. The E-Union server acts as a data centre that collects information from its members and passes information to the required members.

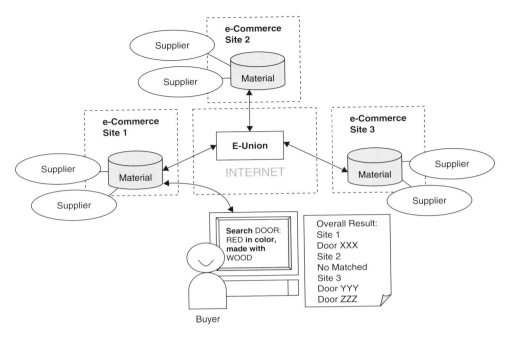

Figure 9.3 Open E-Union framework

Existing e-business systems are developed based on different computational architectures, platforms and software. It is not a trivial task to make them interoperable. There are several issues that need to be addressed for realizing the E-Union concept. The first issue is how an individual e-business system finds services from other sites and communicates with them. The second issue is how one system understands the various types of data from other systems, such as the query, order and product information. The E-Union framework does not require its members to change their existing data structure, as this is very difficult if not impossible. Instead, the E-Union adopts XML as a data representation standard and provides applications to convert its members' data into standard XML documents for intercommunication. The applications for data collection, conversion, discussion and transmission can be developed using different technologies.

9.6 Standardization of construction products information

One of the keys to systems interoperability is the ability to understand meaning of the information provided by each system. Currently e-business systems for construction product procurement present products information in different formats. They provide different interfaces for accessing product information and the types of product information are various in each catalogue. While establishing a standardized construction product information, format can facilitate buyers in product search and selection; it is important to let product information providers keep their own style of product information presentation in electronic catalogues as diversity in product information presentation is considered an important differentiation possibility to attract customers (Keller and Genesereth, 1996).

Standardization of construction product information facilitates information exchange between computers. Search engines cannot present accurate search results if the semantics of information is not standardized. Currently there are two major standards in the building and construction domain that specify the representation of construction product information. They are bcXML developed in the eConstruct project (eConstruct, 2003), and aecXML currently under development by the International Alliance for Interoperability (IAI) in North America (IAI, 2003). The eConstruct project aims to help the European building and construction industry to build faster, cheaper and better by developing a new communication technology specifically tailored to the industry's need. IAI is a global standards-setting organization committed to promoting effective means of exchanging information across all software platforms and applications serving the AEC+FM community by adopting a single Building Information Model (BIM). IAI defines specifications for Industry Foundation Classes (IFC) as BIM. IFC BIM are published in ifcXML and aecXML for e-business and Internet purposes. IFC provides a means to encode and store information for the entire project in a model that can be shared between diverse project participants; aecXML is intended to support specific business-to-business transactions over the Internet. E-Union adopts aecXML as the standard of information representation, as aecXML has schemas for both construction products and business process related schemas, which enables sharing of both product information and business services among the E-Union members.

Both of the above standards specify construction product information using XML. XML is a simple and very flexible text format derived from Standard Generalized Markup Language (SGML) (W3C, 2003). It is a portable and widely supported open technology for describing data, and is becoming the standard for storing data that is exchanged between applications. Construction product information can be described in an XML document that can be read by both humans and machines. Attributes of construction products, such as price, dimension and weight can be represented structurally and hierarchically in the tags of an XML document. Figure 9.4 shows an example of an XML document containing construction product information.

An XML document contains individual units of markup called elements. An element contains one or more pairs of tags and data. The name of a tag in the XML document of Figure 9.4 represents a product attribute. The relationship between attributes is shown by the hierarchical structure of the tags. Data inside a pair of tags is the value of a product attribute.

```xml
<?xml version="1.0" encoding="utf-8" ?>
<product>
    <category>
        <first>Door</first>
        <second>Wood>/second>
        <third>Plain</third>
    </category>
    <dimension>
        <height>2</height>
        <width>1</width>
        <depth>.08</depth>
        <unit>meter</unit>
    </dimension>
    <weight>
        <amount>30</amount>
        <unit>kg</unit>
    </weight>
    <price>
        <amount>800</amount>
        <currency>HKD</currency>
        <unit>per unit</unit>
    </price>
</product>
```

Figure 9.4 An XML document showing construction products information

Processing of XML documents requires a program called an XML parser, which is responsible for checking an XML document's syntax and making the XML document's data available to applications. Some popular parsers include Microsoft's MSXML, the Apache Software Foundation's Xerces and IBM's XML4J. XML documents can reference optional documents that specify how the XML documents should be structured. These optional documents are called Document Type Definitions (DTDs) and Schemas. DTDs provide a means for type checking XML documents and thus confirming that elements contain the proper attributes and arrangement.

9.7 The Web Services model of interoperable construction products catalogues

The use of the Web for application-to-application communication is becoming more and more popular. The programmatic interfaces made available for this use are referred to as Web Services (W3C, 2002). Web Services describe a standardized way of integrating Web-based applications using open standards over an Internet protocol backbone (Webopedia, 2003). The open standards include XML, Simple Object Access Protocol (SOAP), Web Services Description Language (WSDL) and Universal Description, Discovery and Integration (UDDI). Data is tagged in XML and is transferred using SOAP over the Internet. SOAP is a platform-independent protocol that uses XML to make remote procedure calls over HTTP. The

Web Services available are described in WSDL and can be listed in a UDDI directory for other applications to explore them. Two popular platforms for developing and deploying Web Services are Microsoft ASP.Net Platform (Microsoft, 2003) and Sun Java 2 Platform (Sun, 2004). Services previously possible only with the older standardization service known as Electronic Data Interchange (EDI) increasingly are likely to become Web Services (Huang, 2003). Experts and visionaries believe that the benefits of XML Web Services will be instrumental in propelling explosive business growth over the next few years (Yehuda, 2001).

To achieve interoperability between e-business systems for construction products procurement, it is necessary to have a standard interface for data exchange and also a standard method of data representation. Web Services and XML-based standard method of construction product representation together make e-business systems for construction products procurement interoperable. Figure 9.5 shows the Web Services model of interoperable e-business systems for construction products procurement.

The model shown in Figure 9.5 is an application services provider (ASP) solution. Five types of identified practitioners that will provide or retrieve information to and from e-business systems for construction products procurement are agent, manufacturer, supplier, buyer and information provider. Manufacturers provide information on their products in the catalogue. Suppliers provide information on the products that they sell. Agents need to search for products that meet their client's requirement. Information providers provide products information to their subscribed users. Buyers search for products that meet their project requirements. In this Web Services model, the Web Services server links up these practitioners' e-business systems through a common Web Services interface. The server holds a standard format of products information representation which can be bcXML, aecXML or any other standards. All practitioners use this standard for communicating products information.

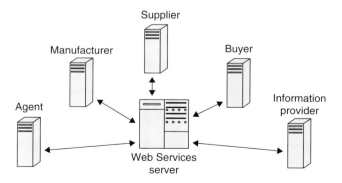

Figure 9.5 Web Services model of interoperable e-business systems for construction products procurement

Under the Web Services model, a practitioner having an e-business system can tell the server the types of products they have and implement Web Services interfaces for communication with the server. Practitioners requiring product information can link to the server and get information from other connected practitioners. The Web Services server acts as a mediator to facilitate sharing of construction product information among the connected practitioners. The server maintains a list of connected practitioners and also the types of product information they provided.

The increased construction product information flow in this Web Services model may alter the supply chain. Relationships between buyers, sellers and agents may change and it may create new channels of distribution. Figure 9.6 depicts the partnership environment in the Web Services model. For instance, a brick manufacturer can form a partnership with a cement manufacturer so that when a buyer searches for product information in the catalogue of the brick manufacturer, he can also get product information from the cement manufacturer, and vice versa. Both manufacturers can suggest buyers to buy products from their partner, which can lead to increased sales for both manufacturers. On the other hand, buyers can get more product information for decision making.

The above partnering strategy will be difficult and costly to achieve without the Web Services environment. Without Web Services, the brick manufacturer has to keep updating the type of products that the cement manufacturer sells. Product information from the cement manufacturer has to be interpreted, rearranged and input into the product catalogue of the brick manufacturer. These processes involve considerable effort, and information in the product catalogue may not be up-to-date due to the time required to process information. However, in the Web Services model, manufacturers do not need to worry about updating product information from their partners as all product information is linked and standardized by the Web Services server.

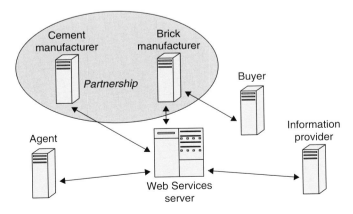

Figure 9.6 Partnership environment in the Web Services model

By enabling product information sharing between the practitioners, buyers and sellers can operate with lower levels of ambiguity and uncertainty due to the provision of greater volumes of timely and accurate information, thereby enabling them to make business decisions in a more efficient and effective manner.

9.8 The E-Union Web Services prototypical implementation

A prototypical implementation of Web Services for enabling interoperability between e-business systems for construction products procurement has been undertaken on a platform named E-Union. E-Union is designed to facilitate information sharing between e-business systems that trade construction products (Kong *et al.*, 2004). Currently E-Union is connected to two systems: a construction products e-trading system named COME (Kong *et al.*, 2001a) and a surplus construction material trading system. Figure 9.7 shows the current system architecture of E-Union with the two e-trading systems.

The E-Union server and the two member servers use Microsoft Internet Information Services as the application server and Microsoft SQL Server 2000 as the database server. Microsoft ASP.Net framework, Microsoft SOAP and Microsoft SQLXML are installed in these servers to provide Web

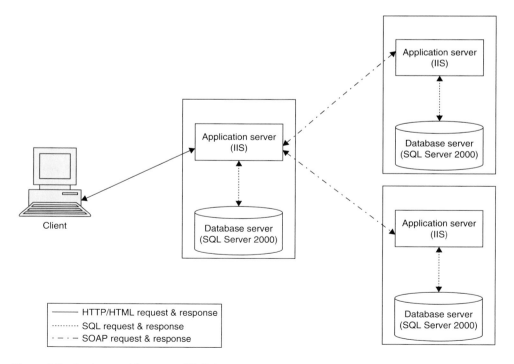

Figure 9.7 System architecture of E-Union

Services. The E-Union server itself does not contain any construction product information. It allows its members to register in it and to specify which types of construction products they have. Basically, E-Union provides services to a client who wants to access product information in E-Union's member Websites. A client firstly accesses the Web-based interfaces of E-Union and then sends a request for certain construction products information through HTTP/HTML. The E-Union application server receives the request and gets information from the database server through an SQL request to find out which member(s) has the required product information. It then gets the required product information from members through SOAP request and sends the information to the client through HTTP/HTML.

9.8.1 Products catalogue searching model of E-Union

The current E-Union system employs a centralized single access product searching model, as shown in Figure 9.8. In this model, the E-Union members can keep their own database structure of product information. The communication interface between E-Union and its members is a standardized Web Services interface. It is the interface between the centralized search engine in E-Union and the data mapping engines in E-Union members. There are two types of communication parameters, query parameters and result parameters. Query parameters specify the kinds of product information that need to be searched and result parameters specify the kinds of product information that are available from members. E-Union members can map their own product information databases with the standardized interface formats. The data mapping method provides high flexibility to

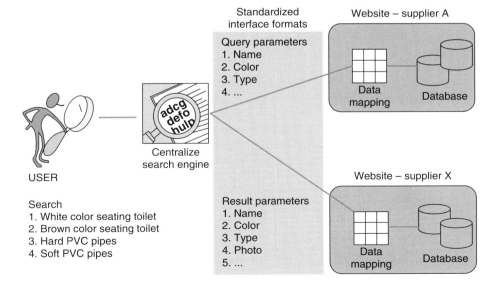

Figure 9.8 Products catalogue searching model of E-Union

E-Union members to maintain their systems. Members only need to update the query and result parameters in data mapping to suit the product information standard of E-Union, rather than putting considerable effort into reconstructing the existing data on product information. An E-Union user can search for construction products through the centralized search engine by inputting different search criteria. The search engine generates requests with standardized query parameters and distributes it to different member Websites. Once a member's Website receives the request, it performs data mapping with the query parameters and databases. The result will then be returned with the standardized result parameters.

9.8.2 Multi-tier products catalogue architecture of E-Union

As discussed in the previous section, a Web-based application typically employs a multi-tier architecture. This is the same in the E-Union products catalogue architecture. The multiple tiers in E-Union are Client tier, E-Union tier and E-Union member tier. Figure 9.9 illustrates this multi-tier architecture.

E-Union member tier

The basic components in E-Union member tier are a middle tier and an information tier, which are in a typical three-tier architecture. The complexity in this tier depends on the member's system architecture. The only requirement is that the middle tier must support Web Services. The Web Server can be Microsoft IIS, Oracle 9iAS, BEA Weblogic and IBM Websphere. The Web Server in the middle tier provides Web Services as the communication bridge between the external request/response and database. The information tier is the database which stores members' products information.

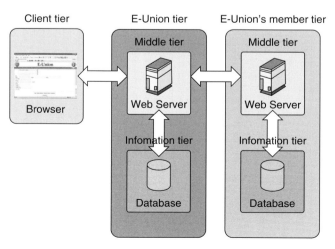

Figure 9.9 Multi-tier architecture of E-Union products catalogue

E-Union tier

Similar to the E-Union member tier, the E-Union tier contains a middle tier and an information tier but it provides different functions. The middle tier under the E-Union tier does not need to support Web Services. It extends its Microsoft IIS Web Server to send SOAP requests and receive SOAP responses to and from E-Union member tiers' Web Services. The Web Server collects search parameters from the client tier and generates multiple SOAP requests to the E-Union member tier. The server waits until all SOAP responses embedded with product information are received. The server then consolidates the results, translates them into HTML format and returns them to the client tier. The information tier is the database containing registered E-Union members' information. Before generating SOAP requests, the Web Server fetches the number and the destination of the members' Web Services from the database. Once the information is fetched, the Web Server knows the number of SOAP requests required to be generated and how to reach the destination members' Web Services.

Client tier

The client tier is some Web-based interfaces in HTML format for interacting with the E-Union tier. These interfaces allow users to specify the types of products they want to search by inputting different product attributes. The product search result will then be displayed in tabular form for easy comparison. Figure 9.10 shows the interface for product searching while Figure 9.11 shows the search results.

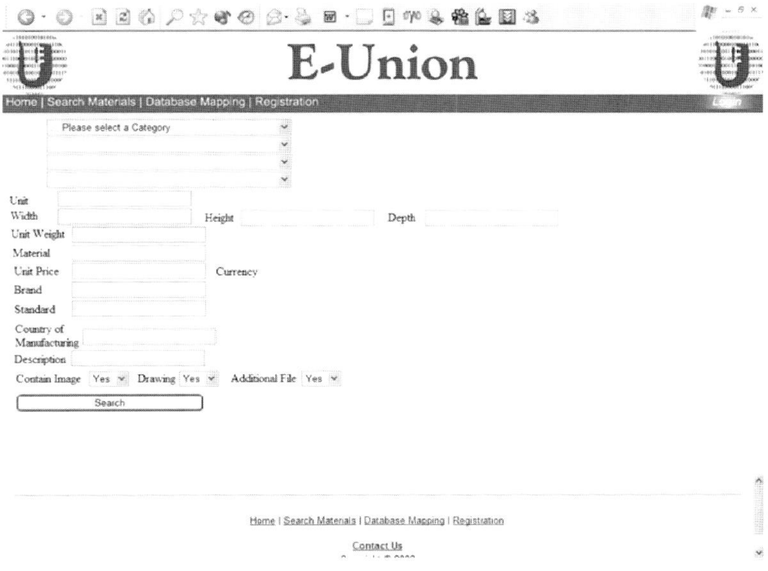

Figure 9.10 The E-Union product searching interface

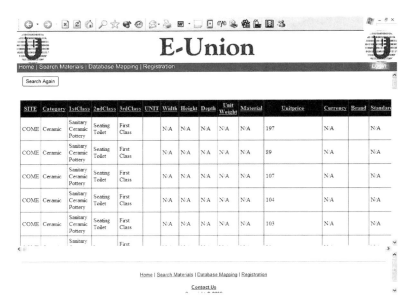

Figure 9.11 The E-Union search result display

9.9 Conclusions

The developments in information technology have changed the way people presently work. Web Services, as an emergent technology for application-to-application communication, has the potential to change the way information is shared and services provided. The increased sharing of standard construction product information in the form of XML documents will ultimately benefit both the buyers and sellers as buyers can be more informed and sellers can increase their market reach. The prototypical implementation of E-Union demonstrated the feasibility of providing interoperable e-business systems for construction product procurement. Currently, the product searching interfaces are only implemented in the E-Union server. These interfaces will be implemented in E-Union members' servers so that buyers can search all E-Union members' product information from any member. There is scope to utilize this interoperable environment to capture more construction products trading-related information and to provide knowledge management functions for assisting business decision making.

References

Agapiou, A., Flanagan, R., Norman, G. and Notman, D. (1998) The changing role of builders merchants in the construction supply chain. *Construction Management and Economics*, **16**(3), 351–361.

Bakos, J.Y. (1991) A strategic analysis of electronic marketplaces. *MIS Quarterly* **15**(3), 295–310.

Castro-Lacouture, D., Harmelink, D. and Skibniewski, M. (2001) Applicability of CAD software enabled with artificial intelligence for drawing management and rebar take-off practices in the US. Unpublished report prepared for VHSoft Software, Hong Kong, Division of Construction Engineering and Management, School of Civil Engineering, Purdue University, May.

EConstruct (2003) *eConstruction: eCommerce and eBusiness in the European Building and Construction Industry: Exploiting the Next Generation Internet*. Available at http://www.econstruct.org/default_frame.htm (accessed 23 February 2004).

Ellsworth, J.H. and Ellsworth, M.V. (1995) *Marketing on the Internet – Multimedia Strategies for the World Wide Web*. Wiley, New York, 225.

Hoffman, D.L., Novak, T.P. and Chatterjee, P. (1995) Commercial scenarios for the Web: opportunities and challenges. *Journal of Computer-Mediated Communication*, **3**(3).

IAI (2003) *Understanding the Different Purposes of IFCs and aecXML in Achieving Interoperability*. Available at http://www.iai-na.org/technical/faqs.php (accessed 23 February 2004).

Keller, A.M. and Genesereth, M.R. (1996) Multivendor catalogs: Smart catalogs and virtual catalogs. *The Journal of Electronic Commerce*, **9**, 259–271.

Keller, A.M. and Genesereth, M.R. (1997) Using infomaster to create a housewares virtual catalogs. *The International Journal of Electronic Commerce and Business Media*, **7**, 41–44.

Kong, C.W. Li, H. and Love, P.E.D. (2001a) An e-business system for construction material procurement. *Construction Innovation*, **1**(1), 43–54.

Kong, S.C.W., Li, H. and Shen, L.Y. (2001b) An Internet-based electronic product catalog of construction material. *Construction Innovation*, **1**(4), 245–257.

Kong, S.C.W., Li, H., Hung, T.P.L., Shi, J.W.Z., Castro-Lacouture, D. and Skibniewski, M. (2004) Enabling information sharing between e-business systems for construction material procurement. *Automation in Construction*, **13**, 261–276.

Li, H., Cao, J., Castro-Lacouture, D. and Skibniewski, M. (2002) A framework for developing a unified B2B E-trading construction marketplace. *Automation in Construction*, **12**, 201–211.

Lincke, D.M. (1998) Evaluating integrated electronic commerce systems. *The International Journal of Electronic Commerce and Business Media*, **8**, 7–11.

Microsoft (2003) ASP.NET. Available at http://www.asp.net (accessed 10 February 2004).

Huang, N.-C. (2003) *A Cross Platform Web Service Implementation Using SOAP*. Unpublished MSc. Thesis, Knowledge Systems Institute.

Sculley, B. and William, W. (2000) B2B exchanges: The killer application in the business-to-business Internet revolution. In: *Proceedings of the 31st Hawaii International Conference on System Sciences (HICSS31)*, Kona, Hawaii, University of Hawaii, January, pp. 149–154.

Stanoevska-Slabeva, K. and Schmid, B. (2000) Internet electronic product catalogues: An approach beyond simple keywords and multimedia. *Computer Networks*, **32**, 701–715.

Sun (2004) *Java Technology and Web Services Overview*. Available at http://java.sun.com/webservices/overview.html (accessed 10 February 2004).

Timm, U. and Rosewitz, M. (1998) Electronic sales assistance for product configuration. In: *Proceedings of the 11th International Bled Electronic Commerce Conference – Electronic Commerce in the Information Society*, Bled, Slovenia, 1, pp. 8–10.

W3C (2002) *Web Services Activity*. Available at http://www.w3.org/2002/ws/ (accessed 10 February 2004).

W3C (2003) *Extensible Markup Language (XML)*. Available at http://www.w3.org/xml/ (accessed 23 February 2004).

Webopedia (2003) *Web Services*. Available at http://www.webopedia.com/term/w/web_services.html (accessed 10 February 2004).

Yehuda Shiran (2001) *Benefits of XML Web Services*. Available at http://www.webreference.com/js/tips/011031.html.

10 Using Next Generation Web Technologies in Construction e-Business

Darshan Ruikar, Chimay J. Anumba and Alistair Duke

10.1 Introduction

Information technology has advanced significantly in recent years and major advances such as the development of innovative computational techniques and communication technologies such as the Internet and wireless technologies are seen as critical to addressing some of the inherent issues associated with construction projects. Construction projects teams usually involve several heterogeneous disciplines working together for transient periods to deliver constructed facilities. Increasingly, the members of these teams are geographically distributed and often cut across organizational boundaries, making collaborative communications difficult. Existing information and communication technologies (ICTs) have done much to address this and provide an appropriate collaboration infrastructure when face-to-face meetings are impossible, expensive, difficult or simply inconvenient. The growing interest in e-business and Web-based collaboration between participants in the construction supply chain is leading to an increase in Web-enabled construction applications. However, these systems still do not adequately meet the requirements of site-based team members and busy mobile project team members. Furthermore, future ICTs need to be able to facilitate the ubiquitous and serendipitous collaboration that construction project teams need for effective project delivery. In this regard, it is important that there is provision for mobile and wireless systems that enable project team members to have access to project information and project participants anytime, anywhere and at the right level of granularity.

From a methodological viewpoint, collaboration support for project partners is currently seen as a 'simple' delivery of the relevant information. Information delivery is mainly static and is not able to take into account the user's changing context. Very often, applications are designed to support a specific process (e.g. project management, quality assurance, health and safety, etc.). Issues associated with user needs, such as the need for dynamic synthesis of contents, provision of context-sensitive and real-time access to multiple information resources has not been adequately addressed.

Emerging technologies such as The Semantic Web, Web Services and Wi-Fi have the potential to overcome the current shortcomings of ICTs in terms of supporting mobile project team members. Collectively, these technologies open up new possibilities for leveraging the capabilities of mobile computing within construction. The Semantic Web's relevance in supporting the mobile construction worker lies in the fact that typical data requirements of construction workers are for either time-critical activities (e.g. Request for Information, collaboration with project partners) or for those that can facilitate task completion (e.g. access to drawings, schedules. etc.). The Semantic Web provides contextual meaning and Web Services technologies allow registration and discovery of services based on the worker's context. Together these technologies have the potential to facilitate construction business processes by allowing mobile workers access to a wide range of data and services on an as-needed basis.

The Semantic Web can help construction organizations to better manage their knowledge this in turn has a positive effect on typical e-business applications such as supply chain management and customer relationship management. This chapter discusses the potential of using the Semantic Web in facilitating construction e-business through the integration of construction services. It initially reviews the technology and then describes the future construction context for its integration and deployment. It reviews the Semantic Web technology as a possible solution to addressing some of the construction industry's fragmentation problems and then illustrates how Semantic Web technologies can offer benefits to construction e-business.

10.2 The construction context

The construction industry is very project-oriented in nature, and it is organized on schemas wherein actors are involved in several projects at the same time (Zarli *et al.*, 2002). Also developing a mechanism to manipulate (capture, store, search, retrieve) knowledge generated during projects has been of interest since the realization that people and knowledge are the most important strategic resources of an organization (Fruchter, 2002). Actors involved in the same project are sometimes thousands of miles apart and practising different working methods pertaining to their respective roles within each project. In addition, most projects can be characterized as virtual organizations that are only established on a temporary contract basis and therefore merely maintain short-term business relationships. All these factors have created the need for extra resources that are required to manage each project and the information generated as a consequence.

Construction projects have long-faced challenges in the areas of infrastructure, logistics and management. The construction industry worldwide has used various methods to try and produce high-quality buildings within the shortest time frame and at reasonable costs. Kalay (2001) states that after years of research it is an accepted fact that efficient collaboration among all of the participating parties throughout the construction process of a particular project is key to the project's success. Effective communication is vital for the success of collaborative and concurrent engineering in construction (Anumba and Evbuomwan, 1999). Knowledge sharing and transfer are two of the main activities in collaboration. Dixon (2000) has shown that the efficiency of collaboration within a project can be improved by the creation of a proper knowledge sharing environment. IT is considered as a popular way to enable efficient information communication and knowledge transfer using media such as the Web. Integrating the existing IT tools with their respective features, such as co-editing, co-browsing and Web conferencing, in order to create an efficient knowledge sharing environment is therefore of great research interest for improving quality in the construction industry (Ingirige et.al., 2002).

A large number of documents and drawings are generated within the design lifecycle of a construction project. Recovering the correct documentation requires that the documents are provided, structured and maintained. The continuous rapid growth in the volume of project information as the project progresses makes it increasingly difficult to find, organize, access and maintain the information required by project users. Also project-related machine stored data which can act as a knowledge resource is no longer contained in one centralized repository but distributed in heterogeneous databases that belong to different individuals, discipline groups, project-teams and organizations. Even though the current information and communication technology (ICT) enables the formation of virtual project teams that can work across geographical and time constraints through virtual workspaces, integrating the heterogeneous information sources (particularly ones that contain weakly structured information) remains a difficult task in the construction sector. The wide use of low-level technologies, mostly adhering to hyperlinks and keyword search which are available independent of a wider knowledge management system (Ding et al., 2003) and lack of meta-level data structures (Christiansson, 2000) is the main reason behind the phenomenon that has led to the development of non-integrated data.

10.3 The need for the Semantic Web

In recent years, the utilization of some Web Services solutions has offered some partial answers to the construction industry's problems. The use of Web Services that can improve the quality of collaboration has motivated the evolution of these technologies in the construction industry since the 1990s. Zhu and Issa (2001) and Christiansson (2000) state that Web Services are a good medium to store, process and manipulate the massive amount of information that is generated during any construction project. The advantage of using Web Services in a construction context is to efficiently communicate the relevant information and knowledge representations fast, regardless of time and geographical location constraints and in a reliable format. A number of construction projects are based on online project collaboration solutions and a new community of virtual construction team members has emerged (Anumba *et al.*, 2003). Several attempts have been initiated for the development of Web-based applications particularly for project management (Stouffs *et al.*, 2002). The aim of these applications is to integrate the developing Web-based technology into construction in order to facilitate the project management process. However, the application of Web Services in the early design stage, such as planning and conceptual design and Web Services that cater to the needs of the on-site and mobile workforce are still rare.

As already mentioned, the growth of the Internet and the phenomenon of globalization has resulted in many organizations and project teams being increasingly geographically dispersed. To cope with this, organizations require knowledge management tools that enable better understanding of the distributed organizational and project-specific digital knowledge and its corresponding containers, thus, enabling efficient collaboration as well as knowledge capture, representation and user adapted access. Most of the currently available knowledge management tools have limitations as described below though they were designed to deal with operations of relevance to the knowledge lifecycle of a particular organization. According to Ding *et al.* (2003) some of the limitations of current information management systems are:

- Information searching is mainly based on keyword search, which may retrieve irrelevant information due to term ambiguity and omit important relevant information when it is stored under different keywords.
- Manual efforts such as browsing and reading remain the main methods to extract relevant information from textual or other representations. The currently available software agents fail to integrate information from different sources.
- Maintaining large repositories of weakly structured information remains a tough and time-consuming task.

One possible method to overcome these limitations is through the development of push-based information delivery services that are complementary to the existing pull-based services that are available to the user (e.g. project extranets).

In a typical knowledge/document sharing environment the end-user has to seek documents to perform tasks. For this, the user may need to share documents, navigate and query the document database as shown in Figure 10.1. Current document sharing systems consist of a set of pull

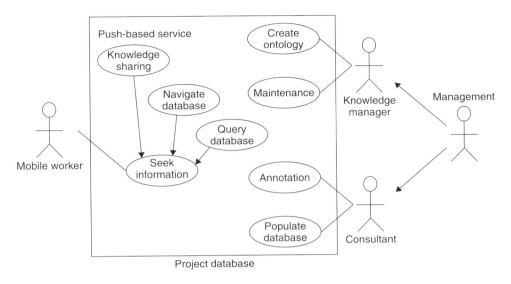

Figure 10.1 Push-based services to facilitate information management

services (i.e. users actively seeking information). What is required besides these pull services are push services to provide the required documentation to the end-user.

Emerging technologies such as the Semantic Web show promise in addressing the current shortcomings related to information management within the construction process. The Semantic Web allows greater access not only to the Web content but also to services on the Web. Using the Semantic Web users and software agents should be able to discover, invoke compose and monitor Web resources offering particular services and having particular properties. Semantic Web increases the utility of Web Services by providing semantics-based brokering capabilities, expressing how terms relate to each other and by enabling dynamic composition of new services. This will enable intelligent access to heterogeneous, distributed information, enabling software products to mediate between user needs and the information sources available. This is particularly important in e-business transactions.

The use of the Semantic Web is not limited to information management on the World Wide Web, it is a technology that will work on internal corporate Intranets. This creates the possibility that the Semantic Web will solve some of the problems associated with current information technology architectures such as information overload, stovepipe systems and poor content aggregation:

- *Information overload*: Information overload is the most obvious problem. Today the Internet has over five billion pages of information and over 300 million users. Thus, the problem has grown worse with the propagation of the Internet, email and instant messaging. There is a huge bias for production of information over reuse. Unless this problem is addressed the problem of information overload will continue to grow.

- *Stovepipe systems*: A stovepipe system is where all the components of the system are hardwired to work together. Thus information only flows in these conduits and cannot be shared with other systems or organizations. For example, a client can communicate with a specific middleware that only understands a single database, which uses a fixed schema. A breakdown of such stovepipe systems needs to occur at all levels of the enterprise information architecture. It is thought that the Semantic Web applications will be most effective in breaking these stovepiped database systems.

- *Poor content aggregation*: Putting together information from disparate sources is a recurring problem in several areas of work for example comparison shopping in e-business, content mining and financial account aggregation. Currently, account aggregation is possible using techniques such as screen scrapping. The main drawback of such a method is that it aggregates HTML data that describes the format of the Web page but does not describe the contents of the Web page. This leads to problems and a time-consuming effort if required data is located in different locations of the screen for different accounts.

There is a possibility that these and other problems associated with current architectures can be solved with the development of the Semantic Web. The next section of this chapter discusses the concept and evolution of the Semantic Web.

10.4 The Semantic Web

10.4.1 Key concepts

Facilities to put machine-understandable data on the Web are becoming a high priority for many communities as is the case with the construction e-business. The World Wide Web Consortium (W3C) has a long-held belief that the Web can reach its full potential only if it becomes a place where data can be shared and processed by automated tools as well as by people. For further development of the Web, future programs must be able to share and process data even when these programs have been designed totally independently. This concept forms the basis of the Semantic Web. The Semantic Web is a vision: the idea of having data on the Web defined and linked in a way that it can be used by machines not just for display purposes, but for automation, integration and reuse of data across various applications (W3C, 2001).

The Internet associated with its most popular application, the Web, provides interconnected infrastructures that are commonly used to facilitate the accessibility of digital resources. However, this Web technology has severe shortcomings that arise from its simple underlying structures and protocols. The current Web works well for posting and rendering all kinds of Web contents but provides very limited support for processing them. This is because most Web contents are stored in natural language chunks, which makes them very much dependent on the human users during search, access, extraction, interpretation and processing. The growing use of Web has also increased the difficulty to manipulate the exponentially increasing amount of information. In response to this, the vision of a Semantic Web was created by Tim Berners-Lee in order to enable automated information access and use based on machine-processable semantics of data. The Semantic Web was defined by him as *'an extension of the current web in which information is given well defined meaning, better enabling computers and people to work in co-operation'* (Berners-Lee et al., 2001). According to the W3C (2001), the Semantic Web is not a separate Web but an extension of the current Web. Using the Semantic Web information is given well-defined meaning, better enabling computers and people to work in cooperation (Berners-Lee et al., 2001).

The Semantic Web should enable greater access not only to content but also to services on the Web. Using Semantic Web technologies, users and software agents should be able to discover, invoke, compose and monitor Web resources offering particular services and having particular properties. As is clear from the above definitions, the concept of the Semantic Web has two broad dimensions:

(1) *Semantics for content determination*: The Semantic Web can be used for corporate knowledge management and making links between corporate information more intelligent this is possible through the development of enterprise level ontologies.
(2) *Semantics for message orientation*: Improving the way information is exchanged and providing a formal data model for Web Services. The Semantic Web should enable users to locate, select, employ, compose and monitor Web-based services automatically. This is possible using Web Services standards such as SOAP, WSDL and UDDI.

According to Tim Berners Lee *et.al.* (2001), 'The essential property of the World Wide Web is its universality. The power of a hypertext link is that "anything can link to anything". Web technology, therefore, must not discriminate between the scribbled draft and the polished performance, between commercial and academic information, or among cultures, languages, media and so on. Information varies along many axes. One of these is the difference between information produced primarily for human consumption and that produced mainly for machines. At one end of the scale we have everything from the five-second TV commercial to poetry. At the other end we have databases, programs and sensor output. To date, the Web has developed most rapidly as a medium of documents for people rather than for data and information that can be processed automatically. The Semantic Web aims to make up for this'.

The W3C describes the Semantic Web as a method for representing data on the World Wide Web. It is a collaborative effort led by W3C with participation from a large number of researchers and industrial partners. It is based on the Resource Description Framework (RDF) (http://www.w3.org/RDF/), which integrates a variety of applications using eXtensible Markup Language (XML) for syntax and URIs (Uniform Resource Indicators) for naming. According to Needleman (2003), the Semantic Web activity in W3C has its roots in the RDF work that W3C performed. RDF, in turn, had its antecedents in some work that was done in the W3C (and outside of it) to develop languages for rating and classifying Web content. The major system that was developed for this purpose was PICS (Platform for Internet Content Selection). The major motivator for the development of PICS was to control undesirable content on the Web.

Like the Internet, the structure of the Semantic Web is as decentralized as possible. Such Web-like systems have generated interest at every level, from major corporations to individual users, and provide benefits that are hard or impossible to predict in advance. Facilities to put machine-understandable data on the Web are becoming a high priority for many communities as is the case with construction e-business. An immediate use of the Semantic Web technology can be seen in Web searches. Using the Semantic Web searches can be made more precise and automated using concepts such as 'partially remembered knowledge'. More generally, the Semantic Web will enable complicated e-business processes and transactions to be carried out automatically. It will allow the processor in a telephone to talk to the processor in a music system and ask it to reduce its volume when the phone rings.

The first steps in weaving the Semantic Web into the structure of the existing Web are already underway. In the near future, the Semantic Web will usher in significant new functionality as machines become much better able to process and understand the data that they merely display at present (Needleman, 2003). To accomplish this vision, efforts to link the existing Web contents to semantic descriptions followed by the creation of a set of applications that can utilize this newly created meta-data are desperately needed and stimulate a new research horizon (Fensel *et al.*, 2002; Semaview, 2002).

10.4.2 Ontologies

Ontologies are decentralized vocabularies of concepts and their relations to which the existing Web contents can refer. These decentralized vocabularies not only define the meaning of Web page contents but also the contents of other information resources, including documents (paper-based) and databases. Ontologies are therefore the kernel of the Semantic Web that allow computers to better categorize, retrieve, query and deduce information from the WWW than the current Web technology (Fensel, 2001; Ding *et al.*, 2003). The concept of ontology applied in artificial-intelligence is to facilitate knowledge sharing and reuse (Fensel, 2001). Ontology is claimed to provide a shared and common understanding of a domain so that people and various application systems can communicate across widely spread heterogeneous sources. As defined by Gruber (Gruber, 1993), an ontology is a formal explicit specification of a shared conceptualization. Thus, it should be machine-readable (Fensel, 2001; Ding *et al.*, 2003). In general, ontology is a graph in which nodes represent concepts or individual objects while arcs represent relationships or associations among concepts. The ontology network takes account of properties and attributes, constraints, functions and rules that govern the behaviour of the concepts (Fensel, 2001). In this respect, ontologies are useful to organize and share information while offering intelligent means for content management as well as enhancing semantic search in distributed and heterogeneous information sources (Fensel *et al.*, 2002; Filos, 2002). In accordance with Maedche (Maedche *et al.*, 2001), establishing domain-specific ontologies is important for the success and proliferation of the Semantic Web.

Figure 10.2 shows the evolution of data fidelity required for semantically aware applications. Instead of just meta-data there is a need for an information stack that comprises of information stacks composed of semantic levels. XML Schema is used in modelling the properties of data classes, as part of capturing and processing meta-data about isolated classes. Level 2 goes beyond data modelling and simple meta-data properties to knowledge modelling. Knowledge modelling enables the development of statements

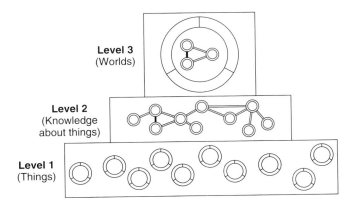

Figure 10. 2 Evolution of data fidelity. (*Source*: Adapted from Fensel, 2003)

between level objects and the way these relationships operate. At Level 3 are the more sophisticated superstructures or 'closed world modelling'; these models are developed using ontologies.

In philosophy, an ontology is a theory about the nature of existence, of what types of things exist; ontology as a discipline studies such theories. Artificial-intelligence and Web researchers have co-opted the term for their words, and for them an ontology is a document or file that formally defines the relations among terms. The most typical kind of ontology for the Web has taxonomy and a set of inference rules. The taxonomy defines classes of objects and relations among them. For example, an address may be defined as a type of location, and city codes may be defined to apply only to locations, and so on. Classes, sub-classes and relations among entities are a very powerful tool for Web use. A large number of relations can be expressed among entities by assigning properties to classes and allowing sub-classes to inherit such properties. If city codes must be of type city and cities generally have Websites, the Website associated with a city code can be listed even if no database links a city code directly to a Website.

With ontology pages on the Web, solutions to terminology (and other) problems begin to emerge. The meaning of terms or XML codes used on a Web page can be defined by pointers from the page to an ontology. Of course, the same problems as before now arise if I point to an ontology that defines addresses as containing a zip code and you point to one that uses postal code. This kind of confusion can be resolved if ontologies (or other Web Services) provide equivalence relations: one or both of our ontologies may contain the information that my zip code is equivalent to your postal code.

Ontologies can enhance the functioning of the Web in many ways. They can be used in a simple fashion to improve the accuracy of Web searches – the search program can look for only those pages that refer to a precise concept instead of all the ones using ambiguous keywords. More advanced applications will use ontologies to relate the information on a page to the associated knowledge structures and inference rules. To date, the applications of ontologies are mainly found in e-business. A number of EU-funded projects have been undertaken to address knowledge technologies in the context of virtual organizations and business collaboration wherein the application of ontologies is the fundamental interest. Amongst them, for example, are ONTOWEB which is a thematic network on ontologies-based information exchange for knowledge management and e-business (On-To-Knowledge, 2002), COMMA which aims at implementing a corporate memory management framework based on agent technologies, and On-To-Knowledge (On-To-Knowledge, 2002) that aims at developing tools and methods for supporting knowledge management relying on sharable and reusable knowledge ontologies. By considering the typical collaboration pattern, it is aware that the construction sector is also committed to virtual-organizational business relationships, which are mainly project-oriented. This resemblance to the e-business sector suggests that the adopted strategy in e-business to improve business collaboration via knowledge sharing is probably applicable to the construction sector.

10.4.3 *Knowledge representation and manipulation*

For the Semantic Web to operate, computers must have access to structured collections of information and sets of inference rules that they can use to conduct automated reasoning. Artificial-intelligence researchers have studied such systems since long before the Web was developed. Knowledge representation, as this technology is often called, is currently in a state comparable to that of hypertext before the advent of the Web: it is clearly a good idea, and some very nice demonstrations exist, but it has not yet changed the world. It contains the seeds of important applications, but to realize its full potential it must be linked into a single global system.

Traditional knowledge-representation systems typically have been centralized, requiring everyone to share exactly the same definition of common concepts such as 'parent' or 'vehicle'. But central control is stifling, and increasing the size and scope of such a system rapidly becomes unmanageable. Adding logic to the Web means to use rules to make inferences, choose courses of action and answer questions. A mixture of mathematical and engineering decisions complicates this task. The logic must be powerful enough to describe complex properties of objects but not so powerful that agents can be tricked by being asked to consider a paradox. Fortunately, a large majority of the information we want to express is along the lines of 'a hex-head bolt is a type of machine bolt', which is readily written in existing languages with a little extra vocabulary.

Two important technologies for developing the Semantic Web are already in place: XML and the RDF. XML lets everyone create their own tags – hidden labels such as that annotate Web pages or sections of text on a page. Scripts, or programs, can make use of these tags in sophisticated ways, but the script writer has to know what the page writer uses each tag for. In short, XML allows users to add arbitrary structure to their documents but says nothing about what the structures mean.

Meaning is expressed by RDF, which encodes it in sets of triples, each triple being rather like the subject, verb and object of an elementary sentence. These triples can be written using XML tags. In RDF, a document makes assertions that particular things (people, Web pages or whatever) have properties (such as 'is a sister of', 'is the author of') with certain values (another person, another Web page). This structure turns out to be a natural way to describe the vast majority of the data processed by machines. Subject and object are each identified by a Universal Resource Identifier (URI), just as used in a link on a Web page. (URLs, Uniform Resource Locators, are the most common type of URI.) The verbs are also identified by URIs, which enables anyone to define a new concept, a new verb, just by defining a URI for it somewhere on the Web.

10.5 Evolution of the Semantic Web in the construction sector

Most of the currently available knowledge management tools used in the construction sector have limitations as described below though they were designed to deal with operations of relevance to the knowledge lifecycle of a particular organization. Information searching is mainly based on keywords search, which may retrieve irrelevant information due to term ambiguity and omit important relevant information when it is stored under different keywords (Ding et al., 2003). Manual efforts such as browsing and reading remain the main methods to extract relevant information from textual or other representations. The currently available software agents fail to integrate information from different sources (Ding et al., 2003). Maintaining large repositories of weakly structured information remains a tough and time-consuming task (Ding et al., 2003).

High degree of automation and scalability in performing tasks with respect to the above-mentioned limitations are expected from exploiting the Semantic Web technologies in the arena of knowledge management. The upscaling of the traditional business arena together with the implementation of efficient support for knowledge management and collaboration puts strains on the existing organizational structure. A leap from the conventional knowledge management system to one built on Semantic Web will facilitate the process (Christiansson, 2000). The Semantic Web-based knowledge management system will be able to keep weakly structured collections consistent, to generate information presentations from semi-structured data and to create semantics of these collections and data which is both machine-accessible and machine-processable (Fensel et al., 2002; Ding et al., 2003).

The concept of using meta-data to reduce the complexity and improve the navigability of data stored in a data repository has been well known since the last decade (Lia et al., 2002). Emerging technologies such as the Semantic Web that use this concept for information storage have the potential to offer better management and distribution of data and knowledge within a project. The use of the Semantic Web allows for better data and information interoperability, fast navigation and mapping. According to Lia et al. (2002) the Semantic Web will enable the development of a simple building core model that forms the basis of a collaborative virtual workspace while handling the redundancy of sub-models that are a nuisance to store, access, retrieve and transfer during the retrieval of any information/data of interest. With this background it can be argued that it is necessary to look into the advantages the Semantic Web has to offer to facilitate the construction process and also to meet the demands of today's mobile workforce. The next section of this chapter will investigate the concept of the Semantic Web and try to explain it with a few scenarios/examples.

Semantic Web technology supports the creation of meta-models (Metamodels.com, 2003). According to Draskic *et al.* (1999) the development of meta-models in the design process has several benefits including reduction in system complexity, provision of model flexibility and integration of multiple and heterogeneous databases. Meta-models also facilitate better interoperability and flexibility of data exchange. It can be argued that Semantic Web supported meta-models can be used in the construction industry mainly to manage the vast amounts of data that is generated during the design and construction process. Making use of the intrinsic characteristic of meta-data to create a self-descriptive meta-model can solve the problem of islands of information that exist within the construction industry. Numerous examples demonstrated that meta-data models might support better data and information interoperability, and also fast navigation, mapping and retrieval of any information/data of interest (Draskic *et al.*, 1999).

To maintain its competitiveness, the construction sector must progress in parallel with the e-business sector to face the challenges of a paradigm shift with respect to the use of innovative ICT including semantic technologies. The main collaboration tools in the construction sector are project extranets (project Websites), workflow management tools and groupware applications for collaborative working. Project extranets build on client-server and Web browser technology to enable distributed project team members to share, view and comment on project-relevant information. This approach is still widely adopted though limitations from its purely document-centric characteristics and limited workflow support have been identified. To overcome the limitations of project extranets, it is necessary to accommodate the increase of information generated throughout the building life, in particular in the early creative design phase wherein fragmented design knowledge capture is of importance.

The use of diverse professional terms impairs communication amongst stakeholders and increases the possibility of misunderstandings. Being aware of the potential impact, the construction sector has undertaken numerous initiatives to broaden the horizon of communication capabilities that are supported by the Internet, and therefore lead to a change of paradigm. Several EC (European Commission) funded projects have been conducted to provide the construction sector with the necessary infrastructure for change. For example, diversity is a project that aims to support and enhance concurrent engineering practices through allowing teams based in different geographical locations to collaboratively design, test and validate shared virtual prototypes (Christiansson *et al.*, 2002). The e-Construct project had the aim of improving Internet-based communication across national and organizational barriers, within the context of e-business. Solutions for transferring and sharing knowledge across ICT systems are therefore the focus of e-Construct. To achieve the objective, a common communication-oriented language, namely bcXML has been defined based on XML with building construction meaning aimed at e-commerce transactions (e-Construct, 2001). e-COGNOS, which aims at offering a generic, modular and open solution for knowledge management in the context of collaboration between actors in a construction project (e-COGNOS, 2001) started in year 2001. More recently, the SEKT project (http://www.sekt-project.com) has been focused on developing semantically enhanced Knowledge Technologies based on semi-automatic techniques for knowledge discovery and ontology generation. These projects promote and demonstrate the evolution tendency from the *document centric* Internet to a *meaning centric* Semantic Web. This shift in focus may meet the requirement of knowledge management practices in the construction sector, which is mostly informal and people-centric wherein abstract concept and meaning are of interest.

The Semantic Web can help organizations including those in construction to better manage their knowledge assets in contrast to the current chaotic method of managing data various business strands that can be supported by better management of corporate memory using Semantic Web applications. This also includes typical e-business applications such as supply chain management and customer relationship management as described in the scenarios later in this chapter. In the context of enabling e-business strategy, the conceptualization of knowledge management is shown in Figure 10.3.

Some examples of the potential uses of the Semantic Web in the construction sector are:

(1) *Knowledge management*: Semantic Web technology can be used for knowledge management with a construction organization that is for building, managing, distributing and evaluating corporate memory within an organization. This Semantic Web consists of Ontologies,

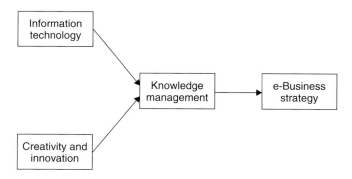

Figure 10.3 **Knowledge management and e-business strategy**

resources (such as documents or persons) and annotations, possibly with modelling of multiple viewpoints.

(2) *Supply chain automation*: By exposing their services in a standard format and through standard data exchange formats (XML) between project partners several tasks can be automated requiring less human intervention.

(3) *Creation of virtual workspaces*: Location and time independent interaction and collaboration among the multidisciplinary partners of a construction project at an early design stage highly influence the quality of final building products. A cross-platform IT supported virtual workspace is therefore necessary. The Semantic Web supports this concept.

(4) *For on-site use*: Semantic Web technologies can be used for the automation of on-site tasks as illustrated by example of project document management and distribution in the next section.

From these examples, those associated with knowledge management and supply chain management are particularly relevant for Semantic Web-based construction e-business, which is discussed in the next section.

10.6 Semantic Web-based construction e-business

The Semantic Web has considerable potential to enhance e-business in the construction sector. This is illustrated here through the detailed description of two construction e-business scenarios.

10.6.1 e-Business Scenario 1: Project document management and distribution using a Semantic Web application

Overview

A large number of documents and drawings are generated within the lifecycle of a construction project. These documents and drawings are not only shared by all project partners (who might be distributed globally) but also internally within the organizations that create them. The rapid growth in the volume of project information as the project progresses makes it increasingly difficult to find, organize, access and maintain the information required by project participants.

The document management procedures for the formal approval and distribution process for project drawings involves many different parties at different locations. Typically consultants (architects, engineers, etc.) prepare documents/drawings and forward these using fax, courier for site usage. Consultants' documents are received at the site office and typically a document controller dates them when received, keeps the originals and issues copies to project team/site team members. On a typical site office drawings will be logged, copied and distributed to the drawing racks kept on-site. As new versions of drawings are issued superseded drawings will be removed from the racks and filed. When work is to be carried out, photocopies of the drawings are given to the operatives and foremen by the site engineers. It is up to the site engineer to ensure that everyone is using the latest versions and that no one is working using out of date drawings. The complexities with this process have become more evident with the advent of design and build contracts where the processes of design and construction run concurrently. The current administration of drawings and documents is mainly paper based and requires a high level of control on the versions of drawings being used on-site. The following drawbacks can be highlighted with the current process:

- Delays in the issue of information/drawings can hamper project progress on site.
- Site contractors are managed by the site engineer who has to keep track of who is using what drawing (i.e. what version) and update them when required.
- Although rare the consequences of using out-of-date drawings if it occurs is very high and very costly. Also time lost on rectifying the situation can affect the project progress.

The use of Web-based collaboration tools have helped automate the process of drawing distribution. However even though they have gone some way to address the complex issue of drawing and document management the use of these systems is still not common on site. The flow of electronic information comes to an abrupt halt when it reaches the construction site. There are two distinct areas in this process that can benefit from the deployment of semantically enabled mobile technologies; the delivery of drawings to the field and the notification of revisions to drawings.

A deployment scenario

In this scenario as shown in Figure 10.4 (envisioned at construction or facilities management stage), the site engineer requires a certain drawing or wants to be informed about an update to a drawing. Using a Web Service client application, mobile construction worker interacts with the project database over the wireless network (e.g. WLAN, GPRS/UMTS). This query is recorded in the project registry. This project database will act as a shared repository for all project-related data for example project documents and drawings which can be accessed by all project partners. Semantic annotation using ontologies developed for all project documents and drawings is envisaged in this scenario. These annotations would facilitate indexing and searching. It would also enable improved ways of information submission and retrieval, by describing resources, and links between them. Such semantic description will also enable

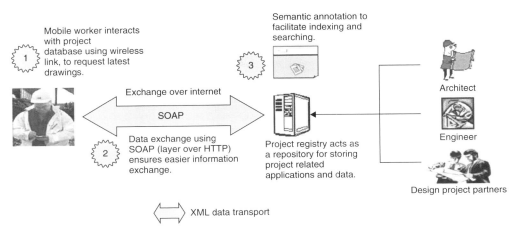

Figure 10.4 Scenario for the deployment of Semantic Web-enabled mobile services

agents to intelligently synthesize the content from multiple information sources, on an *ad hoc* and on demand basis.

To demonstrate the use of the Semantic Web for project data management and distribution in construction OntoShare, a prototype Semantic Web-based knowledge sharing software was used. The OntoShare system was developed as part of the EU project On-To-Knowledge. This system was modified (for data management and ontological structure) and is called OntoWise. OntoWise demonstrated information and data sharing within construction projects using the Semantic Web.

OntoWise: An ontology-based knowledge sharing tool

OntoWise is an ontology-based WWW knowledge sharing environment for a community of practice that models the interests of each user in the form of a user profile. The facilities offered by OntoWise could be used to enhance existing Knowledge Management systems with the direct delivery of information to users, tailored to their specific interests. In OntoWise, user profiles are a set of topics or ontological concepts (represented by RDF classes and declared in RDF(S)) in which the user has expressed an interest. OntoWise has the capability to summarize and extract keywords from WWW pages and other sources of information shared by a user and it then shares this information with other users in the community of practice whose profiles predict interest in the information.

OntoWise is used to store, retrieve, summarize and inform other users about information considered in some sense valuable by an OntoWise user. This information may be from a number of sources: it can be a note typed by the user him/herself; it can be an Intra/Internet page; or it can be copied from another application on the user's computer.

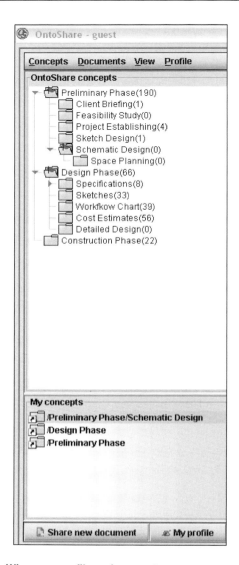

Figure 10.5 OntoWise user profile and concepts

In OntoWise, users can define their profile by subscribing to a set of concepts that are organized in ontology. The membership to a particular concept is shown in the OntoWise user interface by a red flag icon (see Figure 10.5).

OntoWise also modifies a user's profile based on their usage of the system, seeking to refine the profile to better model the user's interests.

When a user finds information of sufficient interest to be shared with their community of practice, a 'share' request is sent to OntoWise via the Java client that forms the interface to the system. OntoWise then invites the user to supply an annotation to be stored with the information. Typically, this might be the reason the information was shared or a comment on the information and can be very useful for other users in deciding which information retrieved from the OntoWise store to access. At this point, the system will also match the content being shared against the concepts (ontological classes) in the community's ontology. Each ontological class is characterized by a set of terms (keywords and phrases) and the shared information is matched against each concept using the vector cosine ranking algorithm. The system then suggests to the sharer a set of concepts to which the information could be assigned. The user is then able to accept the system recommendation or to modify it by suggesting alternative or additional concepts to which the document should be assigned.

When information is shared in this way, OntoWise performs four tasks:

- An abridgement of the information is created, to be held on the user's local OntoWise server. This summary is created using the ViewSum text summarization tool. The summarizer extracts key theme sentences from the document. It is based on the frequency of words and phrases within a document. Access to this locally held summary enables a user to quickly assess the content of a page from a local store before deciding whether to retrieve the (larger amount of) remote information.
- The content of the page is analysed and matched against every user's profile in the community of practice. As when recommending concepts to the user, the vector cosine ranking model is used: here, however, the shared information is matched against the set of terms (words and phrases) created from the union of all terms associated with the concepts to which as user has subscribed (i.e. the concepts which make up the user profile). If the profile and document match strongly enough, OntoWise emails the user, informing him or her of the page that has been shared, by whom and any annotation added by the sharer.
- The information is also matched against the sharer's own profile in the same way. If the profile does not match the information being shared, the system will suggest one or more concepts which strongly match the shared information that the user can then add to their profile. Thus OntoWise has the capability to adaptively learn users' interests by observing user behaviour.
- For each document shared, an instance of the class *Document* is created, with properties holding meta-data including keywords, an abridgement of the document, document title, user annotation, universal resource locator (URL), the sharer's name and date of storage.

A user can select to subscribe to as many concepts as they like at any stage of the hierarchy. They can also select to be informed about information that is added to any sub-concepts of concepts that they have subscribed to. The user's profile is then defined as a list of concepts. As

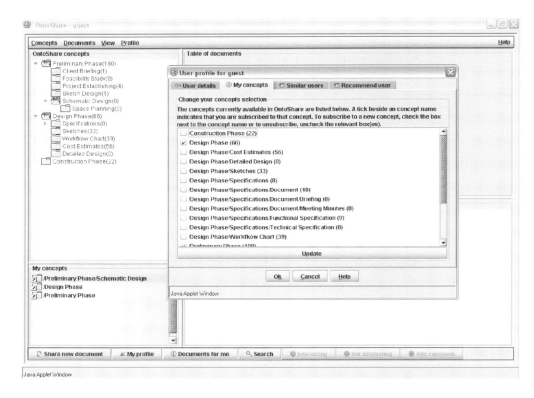

Figure 10.6   Using OntoWise for construction document management

documents are added to the system, the user is only informed if the document is added to a concept to which they have subscribed.

Benefits of using OntoWise

Figure 10.6 illustrates the use of OntoWise for document management and distribution based on the development of a lightweight ontology. Typical using OntoWise based on user profile users can be alerted on any new documents (email), they can search the database for particular documents and they are also presented with personalized information, thus helping provide push-based services to the users.

- *Email notification*: When information is shared in OntoWise, the system checks the profiles of other users in the community of which the user is a member. If the information matches a user's profile sufficiently strongly, an email message is automatically generated and sent to the user concerned, informing the user of the discovery of the information. Thus in cases where a user's profile indicates that they would have a strong interest in information shared, they are immediately and proactively informed about the appearance of the information.

- *Searching the community store: accessing information and people*: Using a button on their OntoWise home page, a user can supply a query in the form of a set of key words and phrases in the way familiar from WWW search engines. OntoWise then retrieves the most closely matching pages held in the OntoWise store, using a vector space matching and scoring algorithm.

  The system then displays a ranked list of links to the pages retrieved and their abridgements, along with the scores of each retrieved page and any annotation made by the original sharer is also shown. Importantly, the user can elect to simultaneously search for other users by selecting the appropriate check box.

- *Personalized information*: A user can also ask OntoWise to display 'Documents for me'. The system then interrogates the OntoWise store and retrieves the most recently stored information. It determines which of these pages best match the user's profile. The user is then presented with a list of links to the most recently shared information, along with a summary, annotations where provided, date of storage, the sharer and an indication of how well the information matches the user's profile (using a thermometer-style icon).

In addition, two buttons are provided so that the user can indicate interest or disinterest in a particular piece of information – this feedback will be used to modify the user's profile. At this point, the system will match the content of the current document against each concept (ontological class) in the community's ontology. As described above, each ontological class is characterized by a set of terms (keywords and phrases) and the shared information is matched against the term set of each concept using a vector ranking algorithm. The system then identifies the set of zero or more concepts that match the information above a given ranking threshold and suggests to the sharer that this set of concepts be added to or removed from their profile in the cases of user interest or disinterest, respectively. The user is then free to accept the system recommendation or to modify it by selecting from the set of suggested concepts.

The OntoWise system has the potential to manage project data efficiently and effectively. The tool helps users to populate the database with documents and drawings and subsequently to enrich the documents with annotations. The application uses ontologies to structure document/knowledge domains, which aim at capturing project/domain data in generic way and provide a commonly agreed understanding of a domain, such that they can be reused and shared across groups.

The development and utilization of next generation Web technology such as the Semantic Web still remains in its embryonic stages and mainly within the research domain within the construction industry. Incremental development and usage of technology can lead to development of tools that support decision making for innovative and routine design in construction by developing an integrated information that is distributed in heterogeneous sources without using one central repository to reduce repetition of workload. It can also aid the development of a knowledge-centric organization.

10.6.2 e-Business Scenario 2: Semantic Web supported KM system for construction collaboration and e-business

The advent of the e-business paradigm has made it necessary for project participants to adopt a systematic approach to organizational knowledge management and to project collaboration. A minimal response is the structured sharing of services and products between employees at the organization level and between project partners at the project level. This is especially applicable in construction as decisions made at the early stage of a product development process, the design process, have severe influences on the quality product (Cohen, 1995). This is a common phenomenon of projects in any domain, including the construction sector (Boverket and BFR, 1994; Formoso *et al.*, 1998) because design is a decision instrument to express product features and production information (Formoso *et al.*, 1998). As already stated in the chapter to improve the design process performance, numerous initiatives have been taken including the partnering concept with its focus on stimulating collaboration amongst the stakeholders from the beginning of a project. Establishing shared value particularly in the context of project-related knowledge tends to improve collaboration amongst stakeholders, and therefore allows them to make fast and accurate decisions at the early stage of design in order to reduce the potential negative costly impact on the later stages. The early stage of a building project is usually referred to as activities that start from client briefing to conceptual design and are inherently iterative. Data and information generated at this stage, such as briefing notes and sketches are mainly informal and not well structured but important to reflect the tacit design knowledge and possibly documented as design rationale. Such weakly structured information is no less important than the structured one such as the final drawings and reports that are generated at the end of every meeting. It is thus an uneasy task to integrate both the weakly- and well-structured information from the perspective of traditional knowledge management (Fensel *et al.*, 2002). The project-related machine stored knowledge is not contained in one centralized repository but distributed in heterogeneous databases that belong to different individuals, discipline groups, project-teams and organizations. Even though the concurrent ICT enables the formation of virtual project teams that can work across geographical and time constraints through virtual workspaces, integrating the heterogeneous information sources particularly ones that contain weakly structured information remains an uneasy task in the construction sector. The widely use of low-level technologies mostly adhering to hyperlinks and keywords search (Ding *et al.*, 2003) and lack of meta-level data structures (Christiansson, 2000) is the main reason behind this non-integrating phenomenon.

The above-delineated shortcomings have motivated the necessity of an innovative knowledge management system. A hypothesis can therefore be formulated that a Semantic Web-based knowledge management strategy is applicable to the construction sector to overcome the dilemma of information and knowledge integration in this domain and furthermore significantly extend the collaboration support amongst project stakeholders. High degree of automation and scalability in performing tasks with respect to the mentioned limitations are expected from exploiting the Semantic Web technologies in the arena of knowledge management. The upscaling of the traditional business arena together with the implementation of efficient support for knowledge management and collaboration puts strains on the existing organizational structure. A leap from the conventional knowledge management system to one built on Semantic Web will facilitate the process (Christiansson, 2000). The Semantic Web-based knowledge management system will be able to keep weakly structured collections consistent, to generate information presentations from semi-structured data, and to create semantics of these collections and data which is both machine-accessible and machine-processable (Fensel *et al.*, 2002; Ding *et al.*, 2003).

The main objective of this knowledge management system is to support decision making in multi-actors environments wherein information is archived in heterogeneous sources. The system is primarily developed to integrate pieces of information generated at the iterative early design stage, to provide fast and precise semantic search, and to capture the intent and rationale behind decisions made particularly during the early design process.

Examination of data in the research domain and industry trends shows that the use of project extranets is the main mechanism to facilitate project information flow in collaborative projects in the United Kingdom. Project extranets are the most popular mechanism used in the case as a means to collect the project-relevant information. However, only structured information, such as design drawings and meeting minutes are available in this information pool. There are severe limitations to using project extranets in regard to the effectiveness of information dissemination. Information is usually categorized based on the preferences of the Web manager, and is archived under different electronic file folders that were created based on the predefined categories. The semi- and unstructured informations, such as briefing notes, design rationale and email messages, are not stored in the project extranets. Email messages are collected in another project-owned digital information source while the paper-based information was kept in personal archives such as file cabinets. Telephones is the most commonly used communication means in the project, that is the case, to share updated information which context was sometimes the intent behind a decision made. Drawings are generated at every stage with respect to the design change, but only the final versions are uploaded to the project Web. Briefly, such descriptions reflected the implication of fragmentary communication and information flow. Apart from that, another significant shortcoming of the project extranets in use is the dependency of inefficient human efforts in processing such as searching, browsing and extracting the archived information.

To cope with the delineated shortcomings, the architecture for an innovative prototype system was built on ontologies that were defined with RDF Schema (RDFS) (W3C, 2001). RDF and its schema are the *de facto* standard for expressing simple meta-data. A number of recently developed and publicly available RDF tools have made the RDFS the core of the prototype system. Scenarios were used in the design of the system architecture and as a basis for future prototype verification through user testing. An example of the scenarios used for designing the system architecture is outlined as the following:

Say an Engineer wants to know who all the Architects involved in Project C. Very quickly, he inputs the name of Project C to access the ontologies (RDFS) built. A list of properties that are defined in the RDFS of Project C is then displayed in the drop-down list box to assist the Architect to search the information of interest. By choosing the property named 'has-role' and filling in the provided dialog boxes to construct a simple query for narrowing down the search scope, the Engineer is then displayed the searched result. The result consists of the names of all the Architects involved, and their profiles.

The excerpt of the RDFS whereof the prototype system is depicted in Figure 10.7. The system is developed with high flexibility so that it can accommodate an unlimited amount of new ontologies in the future. Such ontologies should be defined in a way to capture the domain knowledge. The relevant domain in this case was the project management of a construction project.

With respect to the above scenario, this early prototype system provided functionalities of semantic search of information that was distributed in heterogeneous digital sources. The project-specific ontology consisted of few modular parts, which respectively was ontology, for example the 'team-profile ontology' and the 'early-design-process ontology' that described another aspect of interest, such as the team profiles and the early design process flow, respectively. The project-specific ontologies are linked with each other to provide the expandable capability of the future prototype. The modular characteristic of the ontologies network, which was accessible through the uniquely specified URI, enabled information to be distributed. Such a system allows:

(1) The integration of information that is distributed in heterogeneous sources without using one central repository to reduce repetition of workload.

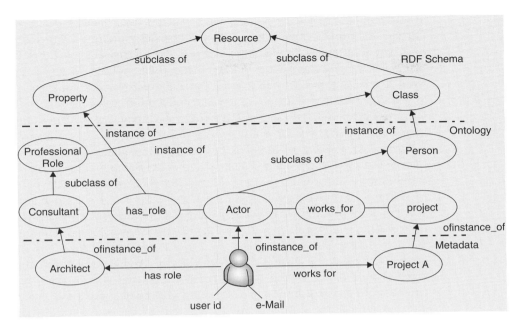

Figure 10.7   Excerpt of RDF(S)-based lightweight ontology

(2) The capture and storage of discussion content wherein design rationale and decision intent are intrinsically encapsulated.
(3) The organization of the captured information in a way that is both human and machine-readable.
(4) The contextualization of the captured information in a representation that may improve the human's efficiency in interpreting its implicit meaning.

Thus, by making use of the technologies of Semantic Web and ontologies, the conventional design documents and drawings can be upgraded to a dynamic and semantically structured medium. The implication is that this medium may handle the mass quantity of design information effectively by eliminating the extra workload, either real or perceived, of having to use extra applications to run a project-oriented knowledge (or information) base, such as the project extranet.

10.7   Summary

Traditional construction knowledge-representation systems are typically centralized, requiring everyone to share exactly the same definition of common concepts. But central control is often domineering, and increasing the size and scope of such a system rapidly becomes unmanageable.

Traditionally knowledge-representation systems also have their own narrow and idiosyncratic set of rules for making inferences about their data. The chapter has introduced how emerging technologies, such as the Semantic Web, can contribute to the creation of a Web of Construction Knowledge and Services that will facilitate construction e-business. Within this Web, construction project team members can have ubiquitous and serendipitous access to project personnel, project information and a vast pool of construction knowledge and services. This will, in future, offer considerable benefits in terms of project management, knowledge management, supply chain management, integration of distributed applications and services and improved construction e-business.

Clearly the future emergence of a Web of Construction Knowledge and Services will result in new e-business processes in the construction industry. Construction researchers will need to seize this opportunity to establish mechanisms and applications that will make the transition from current methods both profitable and smooth. Industry practitioners, on their part, need to be open-minded and adapt to the new ways of working. The potential benefits are considerable but much work needs to be done before they can be realized. Furthermore, the effectiveness of the Semantic Web relies on the development of shared ontologies and semantic standards to ensure increased interoperability across devices, platforms and applications. At present construction enterprises perform their processes in different ways, using different terminologies and modes of operation. Hence the full realization of the vision of Semantic Web-based construction e-business is possible only when the construction industry agrees on common standards and ontologies for process and product description.

References

Anumba, C.J. and Evbuomwan, N.F.O. (1999) A taxonomy for communication facets in concurrent life-cycle design and construction. *Computer-Aided Civil and Infrastructure Engineering*, **14**, Blackwell Publishers, 37–44.

Anumba, C.J. Aziz, Z. and Obonyo, E. (2003) Mobile communications in construction: Trends and prospects. In: *Proceedings of the 2nd International Conference for Innovation in AEC*, Loughborough, UK.

Berners-Lee, T., Handler, J. and Lassila, O. (2001) The semantic web. *Scientific American*, **284**(5), 34–43.

Boverket and BFR (1994) Building & Health, Educational Campaign for Healthy Building, National Board of Housing, Building & Planning (Boverket), Swedish Council for Building Research (BFR), **D1**.

Christiansson, P. (2000) IT in distributed open learning environments. In *Construction Information Technology 2000 – Taking the Construction Industry into the 21st Century* (G. Gudnason ed.). Icelandic Building Research Institute, Reykjavik, Iceland, pp. 197–208.

Christiansson P., Da Dalto Laurent, Skjaerbaek J.O., Soubra S., Marache M., 2002, "Virtual Environments for the AEC sector – The Divercity experience". *ECPPM 2002 Proceedings European Conference of Product and Process Modelling. eWork and eBusiness in AEC*. (Editors: Ziga Turk, Raimar Scherer). Swets & Zeitlinger Publishers, Lisse The Netherlands. ISBN 90 5809 507 X. 9–11 September 2002, Portoroz, Slovenia. (pp. 49–55)

Cohen, L. (1995) *Quality Function Deployment: Deployment: How to Make QFD Work for You*, Addison-Wesley Publishing Company, Reading, MA.

Ding, Y., Fensel, D. and Stork, H.-G. (2003) The semantic web: from concept to percept. *Austrian Artificial Intelligence Journal (ÖGAI)* **24**(3).

Dixon, N.M. (2000) *Common Knowledge: How Companies Thrive by Sharing What They Know*. Harvard Business School Press, Boston, MA, USA.

Draskic, J., Goff, J.-M.L. and Willers, I. (1999) *Using a Meta Model as the Basis for Enterprise-Wide Data*. Available at http://www.computer.org/proceedings/meta/1999/papers/58/rmcclatchey.html.

e-Construct (2001), IST-1999-10303 D103 Final Edition of the bcXML Specification, Available at http://www.econstruct.org/6-Public/bcXML_CD/PublicDeliverables/d103_v2.pdf

e-COGNOS (2001), IST-2000-28671 D2.1 e-COGNOS Base Technology Selection, Available at http://www.e-cognos.org/Downloads/WP2/e-COGNOS%20D2.1.pdf

Fensel, D. (2001) Ontologies: Silver Bullet for Knowledge Management and Electronic Commerce, Springer-Verlag, Berlin.

Fensel, D., Bussler,B., Ding, Y. and Omelayenko, B. (2002) The web service modeling framework WSMF. *Journal of Electronic Commerce Research and Applications*, **1**(2), Elsevier Publications, UK.

Filos E., (2002). European collaborative R&D projects related to the "Smart organization". A first evaluation of activities and implications for construction, *In the Proceedings of the Conference on eWork and eBusiness in AEC, ECPPM 2002 Proceedings European Conference of Product and Process Modelling. eWork and eBusiness in AEC*. (Editors: Ziga Turk, Raimar Scherer). Swets & Zeitlinger Publishers, Lisse The Netherlands. ISBN 90 5809 507 X. 9-11 September 2002, Portoroz, Slovenia. (pp. 27–32)

Formoso, C.T., Tzotzopoulos, P., Jobim, M.S. and Liedtke, R. (1998) Developing a protocol for managing the design process in the building industry, *6th Annual Conference of the International Group for Lean Construction*, Guaruja, SP.

Fruchter, R. (2002) Metaphors for knowledge capture, sharing and reuse. In: *Proceedings of European Conference of Product and Process Modeling (ECPPM) on eWork and eBusiness in AEC* (Z. Turk ed.) September, Portoroz, Slovenia.

Gruber, T.R. (1993) Toward principles for the design of ontologies used for knowledge sharing. *International Journal of Human-Computer Studies*, Special issue on Formal Ontology in Conceptual Analysis and Knowledge Representation. Available as technical report KSL-93-04, Knowledge Systems Laboratory, Stanford University, at: http://ksl.stanford.edu/KSL_Abstracts/KSL-93-04.html.

Ingirige, B., Sexton, M. and Betts, M. (2002) The suitability of IT as a tool to facilitate knowledge sharing in construction alliances. In: *Proceedings of CIB W78 Conference 2002*, Denmark. Available at: http://www.cib-w78-2002.dk/papers/papers/cib02-85.pdf

Kalay, Y. (2001) Enhancing multidisciplinary collaboration through semantically rich representation. *Automation in Construction* (10), 741–755.

Lai, Y-C., Christiansson, P., and Svidt, K. (2002) IT in collaborative building design (IT-CODE). In: *Proceedings of the European Conference on Information and Communication Technology Advances and Innovation in the Knowledge Society*, eSM@RT 2002 in collaboration with CISEMIC 2002 (22–23 November), University of Salford, UK, pp. 323–331, Part A. ()

Maedche, A., Staab, S., Stojanovic, N. and Studer, R. (2001) SEAL – A framework for developing semantic portALs. In: *Proceedings of the 18th British National Conference on Databases*, LNCS 2097, Springer, pp. 1–22.

Metamodels.com (2003)www.metamodels.com.

Needleman (2003) http://www.computer.org/proceedings/meta/1999/papers/58/rmcclatchey.html.

On-To-Knowledge (2002) *IST/1999/10132 D43 Final Report*. Available at http://www.ontoknowledge.org/downl/del43-new.pdf.

Semaview™ Inc. (2002) *Concept to Reality What the Emerging Semantic Web Means to your Business* Available at http://www.semaview.com.

Stouffs, R., Tuncer, B. and Sariyildiz, S. (2002) Empowering individuals to design and building collaboration information spaces. In: *Proceedings of CIB W78 Conference 2002*, Denmark. Available at W3C (1998) http://www.cib-w78-2002.dk/papers/papers/cib02-41.pdf.

W3C (2001) http://www.w3c.org/2001/sw/.

Zarli, A., Rezgui, Y. and Kazi, A.S. (2002) Bridging the European research on ICT in construction: the ICCI cluster project. In *Challenges and Acheivements in E-business and E-work (e2002)* (B. Stanford-Smith, E. Chiozza and M. Edin eds). IOS Press, Prague (Czech Republic), 16–18 October, pp. 1242–1249.

Zhu, Y. and Issa, R.J. (2001) Web-based construction document processing via malleable frame. *Journal of Computing in Civil Engineering*, **15**(3), July, 157–169.

11 Trust in e-Commerce

Zhaomin Ren and Tarek M. Hassan

11.1 Introduction

Trust is a cornerstone of e-commerce. The notion of trust has a long history, various approaches have been developed to foster trust building between business partners. Traditionally, these approaches rely on physical contact and paper-based business processes. Although to a large extent traditional principles for trust building may be still valid in e-commerce, they still face much changing and evolving dynamics. The dynamic modes of e-commerce stemming from globalization and the development of new technologies challenge the traditional trust relationships. The absence of interpersonal physical proximity (e.g. recommendations, letters of credit, background checks, handshakes, body language, face-to-face contact and paper documentation) and the lack of overall control in virtual environments create the perception that business in electronic environment is inherently insecure and cannot be trusted.

In typical Business-to-Customer (B2C) e-markets like eBay, for example, goods and services are traded between anonymous sellers and buyers. The basis of business fully depends on seller's description. As buyers normally do not inspect the item to be traded, trading parties might be tempted to cheat. To manage this kind of risk, e-markets like eBay have developed different trust building procedures such as online ranking and reputation system, third party escrow services and online disputes resolution system. Lack of trust, however, is still the biggest impendent for the development of such e-market places. Problems such as disguised identities, misleading item descriptions, the ineffectiveness of e-market's support to dispute resolution and complex cross-border issues seriously hamper the development of such e-markets.

In the context of Business-to-Business (B2B) environments, trust involves more complex issues. A company conducting online procurement and collaboration or joining virtual enterprises, may face a number of risks, for example, losing confidential information. Lack of confidence on business partners or e-business platforms often stops parties from sharing internal data such as sales reports, production schedules, product designs and logistical details with a supply chain partner. Almost 95% of users have declined to provide private information to Websites at one time or another; 63% of these users indicated this was because they do not trust those collecting the data (Hoffman *et al.*, 1999). The survey conducted by Merz *et al.* (2001) reveals that although digital documents have become part of the normal daily activity for most engineering companies, its legal and contractual regulation is not yet common practice. Legal documents validity and responsibility is entirely attributed to the signed paper copy.

Literature review revealed that very little work has been done on trust in e-commerce. A recent investigation revealed that from 275 published articles only three were related to the topic of trust (Ngai and Wat, 2002). Although the research in this area has increased remarkably in the last few years, most of these studies are either derived from traditional trust building approaches or focusing on only technical issues. With trust research in e-commerce still being in its infancy, most of the current literature revolves around the role of trust and does not offer an insight as to how trust may actually be developed and maintained (Papadopoulou *et al.*, 2001). This chapter aims to provide an overview on this aspect.

11.2 Trust and trust building

Trust is seen as the coordinating mechanism which binds relationships together, provides the necessary flexibility, lowers transaction costs and reduces the complexity of relationships. In particular, trust is a key determining factor for commercial relationships wherever risk, uncertainty, or interdependence exists (McKnight and Chervany, 2001). Given its importance to business relationship, trust has been studied in several disciplines such as social psychology, sociology, economics and marketing. It has also become a major concern in the e-commerce environment over the past decade.

11.2.1 *Trust concept*

Due to the different research perspectives and constructs adopted to conceptualize trust, trust has been defined in various ways. Inconsistent and incomplete conceptualization leads to problems in the development, operation and measurement of trust. Below are a few typical definitions of trust:

- Blau (1964): 'The belief that a party's word or promise is reliable and that a party will fulfil his/her obligations in an exchange relationship'.
- Hosmer (1995): 'The reliance by one party upon a voluntary accepted duty on the part of another party to recognize and protect the rights and interests of all others engaging in a joint endeavour or economic exchange'.

As trust typically provides a solution to the problems caused by social or technical uncertainty, trust is often defined in terms of uncertainties:

- Mayer *et al.* (1995): 'The willingness of a party to be vulnerable to the actions of another party based on the expectation that the other will perform a particular action important to the trustor, irrespective of the ability to monitor or control the other party'.

In an e-commerce context, trust among business partners is more appropriately defined by Hosmer, whilst Blau's definition is more suitable to define users' trust in an e-commerce platform. Schumacher (2005) summarized that most conceptualizations of trust identify several main factors:

- Firstly, trust exists in an environment of uncertainty and risk. If parties could undertake a transaction with complete certainty, then trust would not be required and the concept would be trivial.
- Secondly, trust implies the vulnerability of a partner. The extent of the potential loss due to untrustworthy behaviour is typically much greater than the anticipated gains from honest actions.
- Thirdly, trust describes some degree of predictability. It reflects a prediction of a party's behaviour, and implies that the expectancy that it will perform a specific action is high enough for another party to consider engaging in interaction.
- Fourthly, trust exists in an environment of interdependence and mutuality. The parties to an exchange have to believe that their own objectives cannot be realized without reliance upon the other.
- Finally, trust is inherently a positive and good notion. When people refer to trust, they are making a statement about the likelihood of positive outcomes.

11.2.2 *Trust building issues*

As trust is such a broad concept, and is defined in so many different ways, it is necessary to further explore the nature and characteristics of trust, issues related to trust building and the trust development processes, and these form the theoretical basis of this study.

(1) Blomqvist and Ståhle (2005) summarized that competence, goodwill, behaviour and self-reference are four key components of trust in technology partnerships. Trust is increased by – and decreased by the lack of – evidence of these components in parties' actual behaviour and communication.
  - Competence represents a party's technological capabilities, skills, expertise and know-how. It is a necessary antecedent for trust in the business context, especially in technology partnerships,

where complementary technological knowledge and competencies are a key motivation for partnership formation.
- Goodwill implies a partner's moral responsibility and positive intentions towards the other. Expected goodwill is a necessary and active component for trust in any business partnership formation.
- When the relationship is developing, the actual behaviour (e.g. that the trustee fulfils the positive intentions) enhances trustworthiness. The capability and goodwill dimensions of trust become visible in the behavioural signals of trustworthiness.
- Self-reference means a system's capability of autonomy and dependency. Self-reference is demonstrated by the system's ability to define its own existence, the basic idea for being and doing, values, principles and goals, as well as the ability to form double contingent relationships and run a dialogue. Through self-reference the system becomes aware of its identity and capabilities in relation to others.

(2) McKnight and Chervany (1996) provided a typology of interrelated types of trust constructs that helps to distinguish and capture the conceptual meanings of trust and the key trust construct factors. They identified three major trust types: dispositional, institutional, and trusting beliefs:
- Dispositional trust comes primarily from trait psychology, which says that actions are moulded by certain childhood-derived attributes that become more or less stable over time. It means the extent to which one displays a consistent tendency to be willing to depend on others in general across a broad spectrum of situations and persons. It includes faith in humanity and trusting stance subconstructs. Faith in humanity refers to underlying assumptions about people, while trusting stance is like a personal strategy.
- Institutional trust is derived from sociology, which says that behaviours are situationally constructed. In this paradigm, action is not determined by factors within the person but by the environment or situation. In the e-commerce context, it refers to the legal, regulatory, business, and technical environment perceived to support success. Institution-based trust consists of structural assurance and situational normality:
  - *Structural assurance* means that one believes that protective structures – guarantees, contracts, regulations, promises, legal recourse, processes, or procedures – are in place that are conducive to situational success (Zucker, 1986).
  - *Situational normality* means that one believes that the situation in a venture is normal or favourable or conducive to situational success. Situational normality means that trust is the perception that things in a situation are normal, proper, and customary, however, people do not trust others when they face inexplicable, abnormal situations (Garfinkel, 1963).

- Trusting beliefs means that one believes that the other party has one or more characteristics beneficial to oneself. It reflects the idea that interactions between people, and cognitive-emotional reactions to such interactions, determine behaviour. There are four types of trusting beliefs:
  - *Competence* means that one believes that the other party has the ability or power to do for one what one needs done.
  - *Benevolence* means that one believes that the other party cares about one and is motivated to act in one's interest. A benevolent vendor would not be perceived to act opportunistically by taking advantage of the trustor. Benevolence reflects the specific relationship between trustor and trustee, not trustee kindness to all.
  - *Integrity* means that one believes that the other party makes good-faith agreements, tells the truth, acts ethically, and fulfils promises. This would reflect the belief that the Internet vendor will come through on its promises and ethical obligations, such as to deliver goods or services or to keep private information secure. Thus, integrity is more about the character of the trustee than about the trustor–trustee relationship.
  - *Predictability* means that one believes the other party's actions are consistent enough that one can forecast them in a given situation. Those with high predictability would believe that they can predict the e-commerce partner's future behaviour in a given situation.

McKnight and Chervany's analysis of trust has been strengthened and supported by many other researchers (e.g. Zucker, 1986; Parkhe, 1993a, b; Rousseau *et al.*, 1998; Gambetta, 2000; Korczynski, 2000), though the terms and explanations might differ slightly.

11.3 Trust building in e-commerce

Trust in e-commerce involves two closely related aspects: users' trust to e-commerce platforms and business partners' counter-trust to one another. Compared with trust building in general, business issues such as technology, privacy, transference, and e-business infrastructure have critical impacts on trust building in e-commerce (Dutton, 2005). At a macro level, trust in e-commerce is impacted by the complex issues from three large perspectives: trading partner trust (as between organizations in e-commerce), technology assurances, and social infrastructure (Figure 11.1).

11.3.1 Risks in e-commerce

Lack of trust and consequently barriers to participation in e-commerce activities arise due to uncertainties inherent in the current e-commerce environment. These, uncertainties, in turn, create a perception of increased

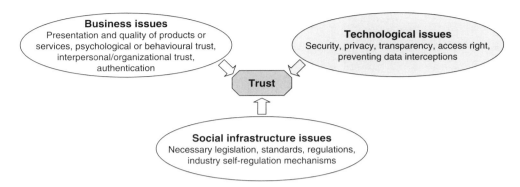

Figure 11.1 Key Factors for Trust Building in E-commerce

Table 11.1 Typical risks in e-commerce

Social infrastructure risk	Technology risk	Business risk

• Lack of e-commerce related laws, regulations and directives (e.g. intelligent property right, professional virtual community, privacy, etc.)
• Inconsistent legislation systems (e.g. cross-border issues)
• Lack of Internet standards
• Non-existing or inadequate user identification systems
• Unclear insurance policy
• Inefficient online dispute resolution system
• Cross-border issues
• Lack of industry self-regulation
• Lack of trusted third party
• etc.
	• Security problems:
	  ○ Transaction security (e.g. online payment, digital signature, etc.)
	  ○ Storage security (e.g. data confidentiality, availability, integrity, access rights)
	• Technical risk (e.g. fraud, virus attacks, technological errors, etc.)
	• Poor designed e-commerce infrastructure (e.g. lack of transparency and traceability)
	• etc.
		• Disguised identity
		• Inaccurate information about business/product
		• Inability or non-willingness to perform
		• Misconception or misleading description
		• Low quality of goods or services
		• Unauthorized copying or use of critical information or digital assets
		• Limited IT knowledge, experience and resources
		• Unclear liabilities
		• etc.

risk, thereby inhibiting the tendency to participate in e-commerce. Uncertainties reduce confidence both in the reliability of B2B transactions transmitted electronically and, more importantly, in the trading parties themselves. Table 11.1 lists some of the risks. Most of them could be a combination of the problems from more than one aspect.

11.3.2 *Trust building issues in e-commerce lifecycle*

Given the complex issues involved in e-commerce, this section will only provide a snapshot on trust building issues in e-commerce. The discussion

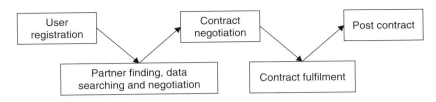

Figure 11.2 e-Business transaction lifecycle

focuses on how an e-market environment, the most popular B2C and B2B platforms, should be developed to build costumers' trust (i.e. trust to e-market, trust to partner and trust to social infrastructure). The analysis is conducted based on Business lifecycle in e-market places (Figure 11.2).

User registration

Registration is one of the important stages for trust building in an e-market. Three kinds of requirements are essential for trust building at this stage:

(1) *Validity of users' general registration information*: e-Markets will request users to provide basic personal/company information for registration. The validity of this information is the basis for e-commerce conducted in e-markets. Therefore, e-markets should be able to check users' key information such as bank account, valid address and communication approaches. For some specialist e-Hubs, financial situation is also requested by the e-markets. A series of trusted financial reference agencies are integrated in such case.

(2) *Validity of users' service related registration information*: Besides the general registration of information, e-markets can also objectively assess the suitability of users (mainly suppliers) for trading purposes. There is a need to enable a user to judge a potential partner's trustworthiness, by developing a cross-industry framework for the verification of prospective partners' trustworthiness, employing advanced business registry systems. This is, however, normally difficult to achieve.

A common approach is that e-markets provide a standard template for suppliers to clearly state their service scope, standards, past experience, project team and finance situation. Depending on the e-market's service level, two kinds of procedure could be adopted:
- Sending a candidate supplier's information to an external expert group to conduct a pre-qualification to the candidate (e.g. e-engineering Hub makes pre-qualification to a new engineering service provider through other consortium members).
- Assigning a legal statement to the candidate supplier to request it to be fully responsible for its service statements (this is adopted by general e-Hubs).

(3) *Service contract* : Service contract is another issue which seriously impacts users' confidence on e-markets. When a user registers in an e-market, the e-market should provide the user with a contractual statement (either in the form of 'Terms and Conditions' or a formal service contract). By agreeing to such a contractual statement, the user enters into contract with the e-market and is legally bound by the terms of the contract. Since the e-market will not be involved in the actual transaction between customers and suppliers, it has no control over the quality, safety or legality of the items advertised, the truth or accuracy of the listings, the ability of service providers or the real requirements of customers. The e-market cannot ensure that customers and suppliers will actually complete their transactions. These legal issues are therefore, extremely important for the e-market to protect itself against any legal actions from users due to the failure of business. To build such a contractual statement, the e-market needs to consider the following issues:

- *Common laws adopted*: The e-market should address the related business laws and Acts essential to the business conducted in it such as Antitrust Acts (e.g. Sherman Act, Clayton Act, Robinson-Patman Act, Federal Trade Commission Act) jurisdiction and enforcement of legal decisions. This will ensure that customers and suppliers will do business in the e-market under the controls of common laws. For example, one cannot avoid the antitrust laws by announcing that the negotiation purpose is to discuss a legitimate item when the substance of the conversation deals with an unlawful exchange of information in the e-market.
- *Contractual laws*: Besides the common laws, contractual laws should specify the business regulation between users and the e-market, and customers with suppliers. Some of the particular contract items should be addressed such as definitions of services, trade rules, liabilities and obligations, limitation and exclusion of liability, software licence, services scope, fees and charges, compliance with laws, warranties, limited warranty and disclaimer, indemnification, termination, copyright, trademarks, confidentiality, data protection, registration information, delay or force majeure, disputes, assignment and notices. Furthermore, ICT-related contractual issues are also essential for the e-market. It should consider issues such as:
  - data storage and protection;
  - confidentiality and access right;
  - legal effects of electronic communication;
  - notification and signature;
  - virus detection and impacts;
  - fraud detection, protection and responsibility.

Partner finding, data searching and service and contract negotiation

At this stage, suppliers advertise their services; customers search for services which they need through the e-market platform. The most important trust issue at this stage is how a party could trust another party during this process. Currently, various trust models have been developed within e-markets to assess users' credibility. It helps users to understand their potential business partners' trustworthiness and decide whether to enter a contract with them or which kinds of procedures to take before entering into the contract. As discussed above, business registries can be understood as a starting point for this work as they provide an effective mechanism for the initial introduction of potential business partners. Major assessments, however, are made by reviewing a user's performance during the business and contract negotiation processes in e-markets and the contract-execution process, which is off e-markets' platforms.

Since an e-market does not hold the liability for the credibility of a user in the e-market, the assessment is mainly done by the user's business partners based on their performance. The e-market is in a position to provide a mechanism to facilitate such an assessment. A general assessment approach could be:

- The e-market provides a feedback forum for users to evaluate the services provided by their business partners. Parameters such as the number of conformities/non-conformities recorded will be defined to demonstrate the user's activities in the e-market.
- A user of the e-market can provide feedback to its business partners based on the business outcome. A simple way of achieving this is by allowing the user to rank the partners (e.g. very good, good, normal, bad, or very bad) with supporting comments. This, however, is only suitable for simple e-market like eBay. For the e-markets involving complex engineering services, users will be required to rank and leave comments based on much thorough criteria provided by the e-market (e.g. service scope, quality, schedule, payment, maintenance, etc.). Such feedback data along with the related business records (e.g. service nature, scope and value) can enable users to establish a baseline of level of trust that is required prior to entering a formal contractual agreement.
- Meanwhile, the party being ranked is also allowed to explain the situation and problems regarding to the rank and comments made by another party.
- Based on the feedback forum, the e-market may also define 'Trust Threshold'. A user may be written-off from the e-market if they get enough negative feedbacks from different partners. Or other users are warned not to deal with those which have gained certain numbers of negative feedbacks and a poor track record.

Apart from trust building models, other major concerns are about data and information exchanges between customers and suppliers (e.g. enquiry, reply, data store, reminders, and negotiation). e-Markets should provide mechanisms to support authorized information exchange, secured data storage, access rights, confidentiality and MoU (Memorandum of Understanding) tracking, and respond to users' reports about misrepresentation, fraud, misbehaviour, virus attacks, and loss of confidentiality. Some of the trust and legal related technical issues include:

- *Digital rights management (DRM)*: Digital rights management restricts the use of digital files in order to protect the interests of copyright holders. DRM technologies should be developed to control file access (number of views, length of views), altering, sharing, copying, printing, and saving. These technologies may be developed within the e-market environment.
- *Intellectual property rights*: Customers and suppliers need to constantly exchange technical information at this stage. Both parties will require assurance that their IPR – inherent in the customer's overall design requirements and the supplier's component technical data, respectively – is respected by the other parties. A legal statement on the ownership and licensed use of IPR by the parties for the purposes of fulfilling their obligations should be defined in the contract.
- *Virtual identity management*: Virtual identity management allows document validation, in such a way that it is possible to guarantee the author identity of a document. This ensures that no changes are made to the original document. A legal statement could be included to protect the authorship of electronic documents.
- *Security*: Users should be confident that public and private data stored in e-markets is well protected and used properly; any transaction made through the e-market is safe; no external attack is caused by e-markets; early warning system to any possible harm to users (mainly regarding to fraudulent emails). Meanwhile, security procedures are also taken to protect e-markets themselves from external threats.

Contract negotiation

For most of the bidding and auction-based B2C e-markets like eBay, the online contracting process is relatively easy. As long as customers and suppliers reach an agreement on prices (may also include payments and deliver methods) within the defined timeline, the contract is signed based on the conditions specified by the e-market. However, for those e-markets which provide comprehensive consultant and engineering services, the online contracting process is much more complex. A major approach to fostering trust building is to provide a reliable contracting environment, which are discussed below.

- *Contract negotiation platform*: e-Markets provide users with an online contract negotiation platform (e.g. the eLEGAL contract editor, URL19

or the e-engineering Hub's OCS, URL5). Both conceptual and technical issues are involved in building such a contract negotiation platform. Also, the design of the platform is highly related to the service content (e.g. what to buy and sell) and users (individuals or enterprises). The following requirements might be necessary:
  - *Authentication of contracting parties*: In order to electronically enter into a contract, the parties will need to be reassured as to the identity of the remote parties. An agreement on authentication mechanisms has to be reached.
  - *Contract negotiation process tracking*: The contract negotiation platform should provide a mechanism to track customers' and suppliers' contract negotiation process (i.e. who proposed, who modified). Customers and suppliers can always check this information during and/or after negotiation, but are not allowed to modify it. The aim is to provide users a transparent and traceable contract negotiation environment. This is important for the sustainable development of the e-market. Technically, this requires the platform to provide functions such as annotation function, smart database, or version control.
- *Contract templates*: Many B2B e-market provides users with various contract templates (or allows customers to adopt their own standard contract templates, or recommend them with special contract template providers). Issues involved in the development of contract templates are the conceptual development of contract templates (which items should be included in a particular contract template) and technical representation of the contract clauses (this also depends on the negotiation platform).

Contract execution and post-contract fulfilment

Although e-markets could end their roles when contracts are signed between customers and suppliers, many e-markets provide post-contract services to their users. Such support mechanisms are often considered as an important part of the e-market services and are important for trust building. These services may include:

- *Facilitating contract-execution related communication*: The e-market may facilitate contract-execution related communication between customers and suppliers (the e-market plays a role similar to a facilitator for a certain period after the contract is signed). The e-market may also play an independent third party's role to record users' communication and discussions related to contract execution (for a certain period).
- *Keeping business and contract negotiation records*: The e-market normally keeps records related to users' business and contract negotiation for a relatively long period (e.g. five years). Users or authorized parties can access these records (but cannot modify them).

Table 11.2 Trust service usage throughout the business process

	Contacting exchange	Finding partners	Credibility checking	Negotiating	Contract signing	Joint business
Authentication	√					√
Authorization	√					√
Assured messaging		√		√		√
Secure storage				√	√	√
Underwriting			√			
Time stamping	√			√	√	
e-Signature	√				√	
Certification/ rating		√	√			√

- *Explaining the contract template clauses and conditions provided by the e-market*: The e-market often needs to explain the contract template clauses and conditions, as requested by users.
- *Recommending legal support or dispute resolution experts*: Some e-markets can also provide supports to users by recommending them legal and dispute resolution consultants.

Similarly, Baldwin *et al.* (2001) summarize the trust service throughout the e-business process as in Table 11.2.

Figures 11.3 and 11.4 illustrate the user's trust building model for e-market at the exploration and contract negotiation stages (for details, see Ren and Hassan, 2005).

11.3.3 Trust building practices in e-commerce

To attract more business opportunities, e-market providers have taken both business approaches and technology approaches to overcome trust barriers. For example, some of the business approaches include:

- Providing unconditional guarantees of safety with an offer to cover any losses due to credit card fraud (e.g. Amazon, URL1).
- Providing detailed explanations of privacy policies on their Websites (e.g. Travelocity, URL2).
- Trying to capitalize on existing brand reputations in the case of established businesses (e.g. Microsoft Expedia, URL3; Barnes and Noble, URL4).
- Building brand recognition for their Web-only businesses (e.g. Travelocity, URL2; Amazon, URL1).
- Building transference-based trust by associating themselves with already-trusted businesses (e.g. e-Engineering Hub, URL5).

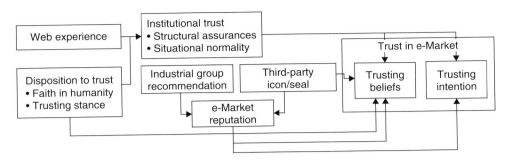

Figure 11.3 Conceptual trust building model for exploration stage. (*Source*: Adapted from Mcknight *et al.*, 1998)

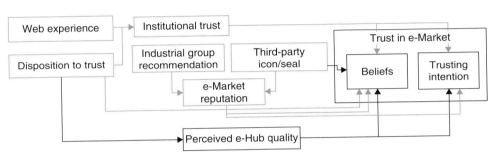

Figure 11.4 Conceptual trust building model for contract negotiation stage. (*Source*: Adapted from Mcknight *et al.*, 1998)

- Including suppliers only approved by trusted third parties, trust seals, trust signals, or intermediaries (e.g. Consumer Reports, URL6; Trust-e, URL7; or CPA society, URL8).
- Building community-based self-regulation systems (e.g. e-Engineering Hub, URL5).
- Providing particular supporting to SMEs (e.g. e-Engineering Hub, URL5).
- Developing online dispute solutions (e.g. Cybersettle, URL9; Web-Mediate, URL10).

Many technologies are also adopted to foster trust building in e-commerce by improving online security and privacy (e.g. confidentiality of sensitive information, integrity of critical information, prevention of unauthorized copying or use of critical information, traceability of digital objects, management of risks to critical information, and authentication of payment information). These technologies include:

- *Identity*: Public key infrastructure (PKI)-based certificate authorities are commonly used to underpin digital identities. Companies such as Verisign (URL11) and ValiCert (URL12) provide mechanisms and services to outsource validation and verification controls.

- *Digital credentials*: Emerging privilege management infrastructures (PMIs) address the very basic need of certifying attributes associated to users and enterprises such as credit card numbers, certified credit limits, ranking information, etc.
- *Recommendation and rating*: Recommendation and rating services are emerging within e-marketplaces to vouch for market participants. e-Market providers are running such recommendation services for participants in their closed communities (e.g. URL13, URL14, etc.).
- *Anonymity*: Anonymity services are becoming popular on the Internet to protect the customers' privacy and security, which is vital for the success of critical e-commerce initiatives. Companies (e.g. Zero-Knowledge Systems, URL15) offer an online anonymity service based on encryption mechanisms and IP masking techniques.
- *Guaranteed message delivery*: Technologies are adopted to assure secured message exchanges between business partners. For example, electronic data interchange (EDI) and WebEDI messaging and business interaction infrastructures provide reliable and trusted infrastructures underpinning interactions within close business communities.
- *Notarization*: Notarization services provide evidence of the existence of documents and messages at particular points in time. For example, Surety (URL16) enables users to notarize electronic files, guaranteeing the content, and enabling the owner to verify the content for a long term. Timestamp.com (URL17) provides a time-stamping service to digitally timestamp digital documents.
- *Storage*: The storage of critical digital documents is extremely important, especially for long period of time, as it is the foundation of accountability. Infrastructure technologies (e.g. Storage Area Networks and Network Attached Storage) provide high availability, replication and survivability of stored documents. Companies like Documentum (URL18) provide solutions to store, index, and manage huge set of documents within enterprises.

11.4 Conclusions

Despite the considerable efforts in both business and technology aspects, lack of trust is still a major barrier for the development of e-commerce. This is because trust in e-commerce involves very complex and dynamic issues, no study cannot address all key aspects of this phenomenon. Most of the studies only touch on a small subset of possible complexities (e.g. technical issues). For e-commerce to reach its full potential, all the impediments to trust must be thoroughly understood and dealt within the design, development and management of e-commerce.

This chapter provides an overall view of trust and trust building issues in e-commerce. It clarifies some fundamental questions of trust such as what trust is, why it is necessary, and how it should be made operational? This provides a basis to evaluate concrete work on operational trust building. It then discusses the uncertainties in e-commerce; and analysed the trust building requirement in e-commerce lifecycle.

Incorporation of the identified trust issues into the system development lifecycle is an area that academics, developers and managers alike, need to consider for future e-commerce developments and innovations. It should be well understood that although there is a broad and deep set of technology problems that need to be addressed, trust building in e-commerce needs a joint effort from social, business, and technology aspects. To be successful, trust-related services must be developed from the three aspects and take account of existing infrastructure. Realizing the full vision will require collaboration amongst many of the likely stakeholders, such as e-commerce platform developers, governments, industry regulators, banks, insurers, lawyers, and users.

References

Baldwin, A., Beres, Y., Mont, M.C., Shiu, S. (2001) *Trust Services: A Trust Infrastructure for E-Commerce (B2B)*. Available at http://hplabs.hp.com/techreports/2001/HPL-2001-198.pdf.

Blau, P. (1964) *Exchange in Power and Social Life*. John Wiley, New York.

Blomqvist, K. and Ståhle, P. (2005) *Trust in Technology Partnerships*. Available at http://www.stahle.fi/material/Trust_in_technology_partnerships.pdf (accessed 25 January 2005).

Dutton, P. (2005) *Trust Issues in E-Commerce*. Available at http://www-sqi.cit.gu.edu.au/WIC2000/docs/Dutton.pdf (accessed 21 January 2005).

Gambetta, D. (2000) Can We Trust Trust? In *Trust: Making and Breaking Cooperative Relations* (D. Gambetta ed). Blackwell, New York, pp. 213–237.

Garfinkel, H. (1963) A conception of, and experiments with, "trust" as a condition of stable concerted actions. In *Motivation and Social Interaction* (O.J. Harvey eds). Ronald Press, New York, pp. 187–238.

Hoffman, D., Novak, P. and Peralta, M. (1999) Building consumer trust online. *Communication of the ACM*, **42**(4), 80–85.

Hosmer, L. (1995) Trust: The connecting link between organisational theory and philosophical ethics. *Academy of Management Review*, **20**(2), 379–403.

Korczynski, M. (2000) The political economy of trust. *Journal of Management Studies*, **37**(1), 1–21.

Mayer, R.C., Davis, J.H. and Schoorman, F.D. (1995) An integrative model of organizational trust. *Academy of Management Review*, **20**(3), 709–734.

McKnight, D.H. and Chervany, N.L. (1996) The meanings of trust. *University of Minnesota MIS Research Center Working Paper Series*. WP96-04, Minneapolis.

McKnight, D.H., Cummings, L.L. and Chervany, N.L. (1998) Initial trust formation in new organizational relationships. *Academy of Management Review*, **23**, 473–490.

McKnight, D.H. and Chervany, N.L. (2001) Conceptualizing trust: A typology and e-commerce customer relationships model. In: *Proceedings of the 34th Annual Hawaii International Conference on Systems Science*, Maui, Hawaii, IEEE Computer Society Press.

Merz, M., Tesei, G., Tanzi, G. and Hassan, T.M. (2001) Electronic contracting in the construction industry. In: *Proceedings of the eBusiness-eWork Conference*, Venice, pp. 595–601.

Ngai, E.W.T. and Wat, F.K.T. (2002) A literature review and classification of electronic commerce research. *Information & Management*, **39**, 415–429.

Papadopoulou, P., Kanellis, P. and Martakos, D. (2001) Investigating trust in e-commerce: A literature review and a model for its formation in customer relationships. In: *The Proceedings of the 7th American Conference on Information Systems*, Boston, MA.

Parkhe, A. (1993a) Strategic alliance structuring: A game theoretic and transaction cost examination of interfirm cooperation. *Academy of Management Journal*, **36**(4), 794–829.

Parkhe, A. (1993b) "Messy" research, methodological predispositions, and theory development in international joint ventures. *Academy of Management Review* **18**(2), 227–268.

Ren, Z. and Hassan, T.M. (2005) Trust and trust building in the single electronic European market. *eAdoption and the Knowledge Economy: Issues, Applications, Case studies (eChallenge 2005)*, Ljubljana, Slovenia.

Schumacher, C. (2005) *The Influence of Trust on SMEs Co-Operation Structuring and Performance*. Available at http://nzae.org.nz/files/%239-SCHUMACHER.PDF (accessed 25 January 05).

URL1: http://www.amazon.com (accessed 21 January 2005).
URL2: http://www.travelocity.co.uk (accessed 21 January 2005).
URL3: http://www.expedia.com (accessed 21 Januay 2005).
URL4: http://www.barnesandnoble.com (accessed 21 January 2005).
URL5: http://www.e-hubs.org (accessed 21 January 2005).
URL6: http://www.consumerreports.org/main/home.jsp (accessed 21 January 2005).
URL7: www.etrust.com (accessed 21 January 2005).
URL8: http://incpas.org/index.asp? (accessed 21 January 2005).
URL9: http://www.cybersettle.com (accessed 21 January 2005).
URL10: http://www.webmediate.com (accessed 21 January 2005).
URL11: http://www.verisign.com.au (accessed 21 January 2005).
URL12: www.valicert.com (accessed 21 January 2005).
URL13: www.clearlybusiness.com (accessed 21 January 2005).
URL14: www.mcphersons.co.uk (accessed 21 January 2005).
URL15: www.zeroknowledge.com (accessed 21 January 2005).
URL16: www.surety.com (accessed 21 January 2005).
URL17: www.timestamp.com (accessed 21 January 2005).
URL18: http://www.e-file-uk.com/ (accessed 21 January 2005).
URL19: http://cic.vtt.fi/projects/elegal/public.html (accessed 21 January 2005).

Zucker, L.G. (1986) Production of trust: Institutional sources of economic Structure. *Research in Organizational Behaviour*, (B.M. Staw and L.L. Cummings eds), **6**, 53–111.

12 Legal Issues in Construction e-Business

Ihab A. Ismail and Vineet R. Kamat

12.1 Introduction

The hype for e-business in the construction industry is now over. With industry players already embracing the change, many are now shifting their focus towards the impact, or non-impact, that e-business has left in their businesses. e-Business applications in the construction industry are poised for restructuring as industry players have recognized that e-business will re-intermediate existing market relations, disrupting some and driving new efficiencies in others, therefore the focus in the next era should on restructuring tasks.

Despite predictions about the success of e-business in construction and how it could change the construction industry (Berning, 2000; Merrill Lynch, 2000; Anumba and Ruikar, 2002), the industry is still struggling with e-business applications. Indeed, the adoption by the construction industry of e-business applications and IT in general has been sluggish but steady (Berning, 2000). The industry should now be concerned with the readiness for the next phase of e-business implementation; a phase that will involve many more players than the innovators and early adopters.

The question of how much value e-business can add to construction operations has been studied in numerous works (Johnson and Clayton, 1998; Mohamed and Stewart, 2003; Love *et al.*, 2004; Stewart and Mohamed, 2004). The need for legally binding e-contracts in construction applications have been noted (Nitithamyong and Skibniewski, 2004). More importantly, the link between the existence of quality legal rules for regulating e-business and the amount of e-business revenue has been established. It has been shown that quality legal rules and enforcement are 'significantly and positively' associated with e-business revenues (Millard, 2000). It is also documented that, if and when, a legally binding e-collaboration and e-business system is established in construction projects, the utilization of such a system increases noticeably (Pena-Mora and Choudary, 2001).

As such, the identification and analysis of e-construction legal risks, coupled with the incorporation of the legal dimension in programming e-business tools can radically improve the utilization of e-business in construction and significantly improve the trust and confidence of the industry in e-business (Ismail and Kamat, 2005). There is a gap, however, in the current literature regarding the specific legal risks related to electronic commerce in construction. This chapter attempts to fill this gap by providing a comprehensive analysis of those legal risks.

12.2 Types of legal risks in construction e-business

The legal risks of conducting business online change every day as new technologies are introduced to the market. These technologies pose important problems to legislators that need quick and effective solutions. The difference between the rates at which e-business technology develops to the rate at which legal framework and rules develop is substantial. Legal risks have not been studied in relation to construction e-business. These risks are identified and discussed in the following sections. The risks are classified under contract formation, validity and errors, jurisdiction, privacy, authentication, attribution, non-repudiation and agency.

12.3 Contract formation, validity and errors

'The law of contracts is broadly speaking, the law of voluntary obligations; that is, the law of obligations that arise because of an express or implied commitment – in particular, a promise' (Fuller and Eisenberg, 2001). According to the Restatement (First) of Contracts 'A promise may be defined as an expression of an intention to act (or refrain from acting) in a specific way, so made as to justify the person to whom the expression is addressed in understanding that a commitment has been made to that person'. In a commercial context, the promises are often exchanged in the form of an offer and an acceptance of the offer. The offer and acceptance, coupled with a valid consideration and mutual assent would, subject to certain limitations, constitute a valid contract.

For example, in the context of construction contracting, the contracting relationship is first initiated by the owner issuing a Request for Proposals (RFP). The RFP is not an offer from the owner to general contractors; the RFP is advertising for general contractors to submit proposals for consideration. The general contractors will study the project based on the RFP information and will submit priced proposals with certain terms and conditions to the owner. The priced proposals act as an offer to enter into an agreement on definite terms and price. Upon selection of the lowest responsible bidder, the owner will notify the successful contractor of its selection. This notification will constitute the legal acceptance of the contractor's offer. Provided there is mutual assent and valid consideration, the owner and contractor will have entered into a legal contract.

In e-business transactions, however, it may not be easy to distinguish the offer maker from the offer taker. This is very critical because a contract is not valid until an offer is accepted and the acceptance is communicated to the offer maker. A recent legal case in the United Kingdom is a good example of this issue. An e-merchant posted a £3.99 price tag for a TV on their Web page instead of £399.00 because of a computer error. In excess of 20 000 TVs were sold for £3.99 before the error was realized. The e-merchant closed the site and a dispute between the e-merchant and the buyers is yet to be settled. The buyers wanted to claim the products they bought but the seller was not willing to entertain the buyers' requests. The key issue is whether the Web page posting of the TV was an offer from the e-merchant that was accepted by the buyers and is thus legally binding, or the Web page posting was an invitation to bid or mere advertising that did not constitute an offer. If the latter is true, then the buyers were the ones who submitted an offer for £3.99; where the seller retains his right to reject the offer and not enter into an agreement (Gallagher, 2000; Pacini *et al.*, 2002).

12.4 Jurisdiction

Jurisdiction is a legal term describing which law is in effect at a given period of time and which court's decisions will be legally binding. Jurisdiction issues arise when parties dispute a contract and try to decide which court has jurisdiction over it. The problem is exacerbated by the fact that in an e-business contract, the question of where the contract was formed becomes daunting. The outcome of the dispute can differ materially if judged under a different set of rules, regulations and laws (Rowe, 1998).

The global reach of the Internet adds yet another layer of complexity: determining jurisdiction over contracts with international elements. The key issue of jurisdiction can be simply explained by comparing the spatial distinction of the legal world to the borderless nature of the Internet. The Internet empowers people to engage in e-business activities regardless of geographical boundaries. In contrast, most of the laws governing contractual relationships, specifically those relating to Construction Law, are limited in one way or another to a certain geographical boundary.

One risk arising from the issue of jurisdiction is unanticipated foreign lawsuits. Foreign in this context means a different state or country. Although the laws regulating e-business vary, the general opinion of the courts imply that companies engaged in activities or online advertising may have to defend lawsuits in different jurisdictions if those activities violate the local laws (Thelen Reid & Priest LLP, 1997). This may very well affect construction companies participating in e-business activities over the Internet, especially in areas such as intellectual property rights and distribution rights.

Companies with local or regional trademarks may find themselves infringing upon others' trademarks when they advertise their products online. Advertising products on the Internet or engaging in e-business over the Internet makes the product accessible globally and is not limited geographically, as trademarks are (Thelen Reid & Priest LLP, 1999b). Similarly, there is a distinction between geographically based (local and regional) distribution rights and distribution over the Internet through e-business activities. Distribution of construction materials or products may have only local or regional distribution rights. Would e-business activity constitute a breach if the distribution rights of both the seller and the buyer are within the geographical boundaries? How about if the seller is within the geographical boundaries of the distribution rights and the buyer is outside; or vice versa?

Jurisdiction risks can also appear in a different form: a company can fail to protect its legal rights due to lack of familiarity with a foreign jurisdiction's procedures or, in case of a dispute, where the parties disputing cannot agree on which court has jurisdiction. This is illustrated in the following example:

Consider the case of an Architect, working out of California, employing an Engineering Consulting firm, working out of New York, for the design of a specialty component of a project in Michigan. The Architect and Consulting firm engage in an e-contract and they never physically meet; collaboration and submission of work is completed online. During construction it was discovered that the Consulting firm's design was faulty and did not meet the Michigan code. The Architect deducts the value of the change from the firm's pay, the firm is opposed to this and stops working; a dispute arises. The question is: which state law has jurisdiction over this dispute? The issue of jurisdiction in this example may be even more challenging if the error was discovered after construction was complete and the different states involved have differing statutes of limitations and/or statutes of repose. Enabled by online project collaboration systems, and e-business tools, the probability of occurrence of this risk is on the rise. The situation is further compounded on global construction projects that involve team members from several countries.

Some of the jurisdiction risks can be avoided by adding choice of law, choice of forum, or arbitration clauses (Thelen Reid & Priest LLP, 1999b). Many risks, however, still exist even when such clauses are made part of the agreement. It is also important to note that not all choice of law and choice of forum provisions are enforced by courts (Gallagher, 2000). Is it a possibility that e-business solutions could be programmed to include those provisions and intelligent agents could be made to distinguish between those provisions that are enforceable and those that are not?

12.5 Privacy

The collection of information and data about people is a characteristic of e-business transactions. e-Business marketplaces and portals derive value from data mining activities (Millard, 2000; Crichard, 2003). Construction companies engaged in e-business will have to manage multiple risks relating to privacy; some relate to personal privacy concerns about their own firms, and others relate to concerns about infringement.

Construction companies wanting to engage in e-business transactions have to give away some information about the company for the purposes of authentication, and attribution; the question is how much information? Online portals and marketplaces collect more information than is needed for legally authenticating an e-contracting party (Smith and Clarke, 2000). They collect information through user registration forms, cookies and track companies' 'bidding behavior'. Who owns this information? Does the information collection process violate any privacy laws? This data mining operation presents a potential risk for construction companies engaging in e-business.

For risks relating to infringement, consider the case of a contractor engaging in e-procurement activities. The contractor keeps information about potential suppliers on the company IT system. It legitimately uses this information to select appropriate bidders. To aid future estimating operations, the contractor records all historical bids on all suppliers who previously electronically bid for projects on its e-procurement system. The historical database includes supplier information, historical bid prices, financial performance collected from pre-qualification forms, safety performance, comments about the supplier performance on previous projects, amongst other information. As a result of a security breach, or a virus, the contractor realizes that the information ended up with a third party – other than the contractor and suppliers. The contractor is liable to the supplier for the loss of information and distribution to other parties and the damages to the supplier that could ensue as a result of this loss.

Another variation to the same risk relates to using online marketplaces. An owner or a contractor places a Request for Proposals on one of the online marketplaces owned by a third party. Suppliers bid on the project after entering information about their respective companies. The third party, owner and operator or the marketplace, have a security breach in which other people had access to stored information about this project mistakenly. Though common sense indicates that the third party should be liable, the contractor or owner may realize that they signed a 'click-through' agreement that limits the third party's liability to such instances; a bad situation to be in.

12.6 Authentication, attribution and non-repudiation

Risks related to electronic authentication, attribution and message integrity are of particular importance to the construction industry. The drive to authenticate e-business buyers and sellers and attribute contracting actions to the proper buyer or seller is in direct conflict with privacy laws (Smith and Clarke, 2000). Attributing an electronic message for an offer or acceptance of an e-contract to the person who purports to send it is yet another risk (Pacini *et al.*, 2002).

The Uniform Electronic Transaction Act (UETA) addresses this risk from a legal perspective by requiring certain authentication levels to ensure proper authentication and attribution, and protecting e-business participants from the actions of hackers (Belgum, 1999; Moreau, 1999; Thelen Reid & Priest LLP, 1999a; Pacini *et al.*, 2002). It is outside the scope of this chapter to discuss the types of non-cryptographic and cryptographic signatures (Hernando, 2003) suitable for e-business transactions. It is important, however, from legal programming and user perspectives to incorporate this signature requirement and comply with it when engaging in e-business transactions.

12.7 Agency

Software agents present the biggest challenge to the current legal system. Legal risks posed by software agents are perhaps the only set of risks that are truly exclusive to the e-business environment. What distinguishes software agents from other software is its degree of autonomy. Agents control their decisions; they learn and act upon their perception of their environment to maximize the goals of its user or programmer (Dzeng and Lin, 2004; Lee, 2004; Ren and Anumba, 2004). Software agents are computer programs that possess a learning capability and can take decisions on behalf of their users and programmers (Dzeng and Lin, 2004; Ren and Anumba, 2004). They act on behalf of their owners to promote the owners' desires, unlike support software that supports the owner in making a decision but leaves the decision for the owner to make (Schoop *et al.*, 2003; Ren and Anumba, 2004). To understand what software agents are, consider the following hypothetical case:

12.7.1 Prelude

Supplier 1 (S1) is responding to an online invitation to bid from the General Contractor (GC) to supply materials for a construction project. The General Contractor is considering two other short-listed suppliers: Supplier 2 (S2) and Supplier 3 (S3). Negotiation is done in a pure e-business setting with each party being represented by a negotiation agent: GC representing the General Contractor, S1, S2 and S3 for Supplier 1, Supplier 2 and Supplier 3, respectively. Each party feeds their negotiation agent with their payoff settings. Payoff settings may include items such as price, delivery dates, warranty terms, payment terms and other special terms.

12.7.2 Agent negotiation

The process of negotiation between agents is initiated by GC. GC receives an offer from S1, S2 and S3 concurrently and evaluates them. GC will then start a negotiation algorithm with the three supplier agents that will try to maximize the value for GC. On the other hand, S1, S2 and S3 will engage in the same negotiation process with GC, each with the goal of maximizing their own benefit. Neither party knows the value curve of any of the others. A critical distinction between software agents and standard software is that agents can learn during this process. More importantly, agents make decisions on behalf of their users without delegating the decision back to them. GC, based on the negotiation with S1, S2 and S3, will analyze bids, negotiate favorable terms, and finalize the transaction with the winning supplier: in this example assume S1.

12.7.3 Variation

In this example, user of agent GC initiated the negotiation. In other examples, GC can automatically initiate the negotiation and make decisions on behalf of its user. Consider that agent GC is linked to an Enterprise Resource Planning (ERP) system. Agent GC will automatically sense that the level of inventory of a certain material is below what is needed to finish scheduled activities on time. Agent GC will automatically initiate a bid by sending an invitation to bid. Agent GC will search for suppliers selling the required material and will evaluate their reputation, proximity to the project location, history of dealings with GC, material availability, etc. Agent GC will then select S1, S2 and S3 based on this search and send them an invitation to bid. The rest of the example will follow until agent GC automatically contracts with agent S1.

Agents with the ability to autonomously initiate actions on behalf of their users are called Initiator agents. On the other hand, agents that mediate an agreement and make decisions on behalf of their users only when a request to make a decision is initiated by the user are called Mediator Agents (Bain and Subirana, 2003b). This distinction is important from a legal perspective. Initiator agents have a higher degree of autonomy; thus posing more challenges to the legal system.

Intelligent software agents exist (Liang and Huang, 2000; Bain and Subirana, 2003b; Dzeng and Lin, 2004; Ren and Anumba, 2004). Few software agents are developed, or are currently being developed, that target the construction industry. (Anumba *et al.*, 2003; Dzeng and Lin, 2004; Ren and Anumba, 2004; Tah, 2004) Many intelligent agents target other industries that can easily be tailored to fit construction (Karageorgos *et al.*, 2003). The hypothesized example is not far from being a reality. C-negotiators is an example of software agents specifically designed and programmed for online e-business in construction. It is an autonomous software agent that can evaluate and negotiate bids automatically with other e-business bidding software agents, or another version of the same agent utilized by different user (Dzeng and Lin, 2004).

Intelligent agents can help solve some of the legal issues that e-business raises. Time-bound negotiation agents, for example, can control option contracts (Lee *et al.*, 2000). Intelligent agents, however, can also pose serious legal challenges to the current legal system and the application of e-business in the construction industry. Agent e-business presents multiple legal risks to the construction industry; some are particular to these types of transactions, while others are similar to the risks associated with e-business contracting. Common risks in e-business contracting, such as jurisdiction, contract formation and errors are discussed elsewhere in this chapter. This section will focus only on the risks that are particular to intelligent agent e-business.

At a conceptual level, the basic challenge about intelligent agents relates to their legal personality: do intelligent agents have a separate legal personality from their owner? Could the Agency Law be applied towards intelligent agents' actions? Can intelligent agents enter into an enforceable contract? Would they be liable for mistakes? How would risk be distributed amongst contracting agents in case of an error? Intelligent software agents' capacity to take actions on behalf of their user is part of their programming and learning. Do they have the same capacity from a legal point of view?

The fact that intelligent agents have the programmed-in capability of making decisions (on behalf of the owner) that may offer better judgment than that of the actual owner, does not necessitate that the law will acknowledge this as contractual consent (Bain and Subirana, 2003c). The answers to those questions are not readily available in Agency Law. The basic premise of Agency Law is that the principal and agent are two separate individuals (Bain and Subirana, 2003a; Bain and Subirana, 2003c).

12.8 Conclusions

Despite the importance of legal issues in the construction industry, the study of legal issues in construction e-business has not been well integrated with the new advances in information technology and e-business. One of the leading efforts towards this integration in the European Union is the eLEGAL project which defined 'a framework for legal conditions and contracts regarding the use of ICT in project business' (e-LEGAL, 2002).

Identifying the legal risks for doing e-business in construction; however, is not enough. It is important that this knowledge is integrated in the development of new applications, business models of services supporting the construction e-business. For example, by learning about issues in contract formation and the differentiation between offer and acceptance, the developers of construction e-business platforms can address these issues in future development of their respective system. Similarly, learning about Jurisdictional issues with e-business contracts would allow contracting parties to, possibly, include provision for choice of law provision in their contract.

Although professionals and academics working in areas relating to the development and support of e-business tools in construction are not required to be legal experts, they should possess enough knowledge relating to the legal risks that exist within this scientific domain and attempt to minimize those risks throughout their work.

References

Aguilar-Saven, R.S.R.S. (2004) Business process modelling: Review and framework. *International Journal of Production Economics*, **90**(2), 129–149.

Anumba, C.J. and Ruikar, K. (2002) Electronic commerce in construction – trends and prospects. *Automation in Construction*, **11**(3), 265–275.

Anumba, C.J., Ren, Z., Thorpe, A., Ugwu, O.O. and Newnham, L. (2003) Negotiation within a multi-agent system for the collaborative design of light industrial buildings. *Advances in Engineering Software*, **34**(7), 389–401.

Bain, M. and Subirana, B. (2003a) E-business oriented software agents: Some legal challenges of advertising and semi-autonomous contracting agents. *Computer Law & Security Report*, **19**(4), 282–288.

Bain, M. and Subirana, B. (2003b) E-business oriented software agents: Towards legal programming: a legal analysis of ecommerce and personal assistant agents using a process/IT view of the firm. *Computer Law & Security Report*, **19**(3), 201–211.

Bain, M. and Subirana, B. (2003c) Legalising autonomous shopping agent processes. *Computer Law & Security Report*, **19**(5), 375–387.

Berning, P.W. (2003) *E-business and the Construction Industry: User Viewpoints, New Concerns, Legal Updates on Project Web Sites, Online Bidding and Web-Based Purchasing*. Available at http://www.constructionweblinks.com/Resources/Industry_Reports_Newsletters/Dec_22_2003/e_commerce.htm (accessed at 11 January 2004).

Berning, P.W. (2000) *E-business and the Construction Industry: The Revolution Is Here*. Avaialble at http://www.constructionweblinks.com/Resources/Industry_Reports_Newsletters/Oct_2_2000/e-business.htm (accessed at 28 October 2004).

Belgum, K.D (1999) *Legal Issues in Contracting on the Internet*. Available at http://www. constructionweblinks.com/Resources/Industry_Reports_Newsletters/June_1999/june_1999.html (accessed 28 October 2004).

Crichard, M. (2003) Privacy and electronic communications. *Computer Law & Security Report*, **19**(4), 299–303.

Dzeng, R. and Lin, Y. (2004) Intelligent agents for supporting construction procurement negotiation. *Expert Systems with Applications*, **27**(1), 107–119.

eLEGAL (2002) *European project IST-1999-20570 2000/11–2002/11*. http://cic.vtt.fi/projects/elegal/public.html.

Fuller, L.L. and Eisenberg, M.A. (2001) *Basic Contract Law*. West Group, St. Paul, Minnesota.

Gallagher, S. (2000) Contracting in cyberspace – A minefield for the unwary. *Computer Law & Security Report*, **16**(2), 101–104.

Hernando, P.D.I. (2003) Electronic commerce. *Computer Law & Security Report*, **19**(5), 363–374.

Ismail, I.A. and Kamat, V.R. (2005) Legal risk analysis, modeling, and programming for e-business in construction. In: *Proceedings of the 2005 ASCE Construction Research Congress*, ASCE, Reston, VA.

Johnson, R.E. and Clayton, M.J. (1998) The impact of information technology in design and construction: the owner's perspective. *Automation in Construction*, **8**(1), 3–14.

Karageorgos, A., Mehandjiev, N., Weichhart, G. and Hammerle, A. (2003) Agent-based optimisation of logistics and production planning. *Engineering Applications of Artificial Intelligence*, **16**(4), 335–348.

Lee, K.J., Chang, Y.S. and Lee, J.K. (2000) Time-bound negotiation framework for electronic commerce agents. *Decision Support Systems*, **28**(4), 319–331.

Lee, W. (2004) Towards agent-based decision making in the electronic marketplace: interactive recommendation and automated negotiation. *Expert Systems with Applications*, **27**(4), 665–679.

Liang, T. and Huang, J. (2000) A framework for applying intelligent agents to support electronic trading. *Decision Support Systems*, **28**(4), 305–317.

Love, P.E.D., Irani, Z. and Edwards, D.J. (2004) Industry-centric benchmarking of information technology benefits, costs and risks for small-to-medium sized enterprises in construction. *Automation in Construction*, **13**(4), 507–524.

Merrill, L. (2000) *The B2B Market Maker Book*. Report No. 5078, Merrill Lynch & Co., New York.

Millard, C. (2000) Four key challenges for Internet and e-business lawyers. *Computer Law & Security Report*, **16**(2), 75–77.

Mohamed, S. and Stewart, R.A. (2003) An empirical investigation of users' perceptions of web-based communication on a construction project. *Automation in Construction*, **12**(1), 43–53.

Moreau, T. (1999) The emergence of a legal framework for electronic transactions. Computer & Security, **18**(5), 423–428.

Nitithamyong, P. and Skibniewski, M.J. (2004) Web-based construction project management systems: how to make them successful? *Automation in Construction*, **13**(4), 491–506.

Pacini, C., Andrews, C. and Hillison, W. (2002) To agree or not to agree: Legal issues in online contracting. *Business Horizon*, **45**(1), 43–52.

Pena-Mora, F. and Choudary, K.K. (2001) Web-centric framework for secure and legally binding electronic transactions in large-scale A/E/C projects. *Journal of Computing in Civil Engineering*, **15**(4), ASCE, Reston, VA, 248–258.

Ren, Z. and Anumba, C.J. (2004) Multi-agent systems in construction-state of the art and prospects. *Automation in Construction*, **13**(3), 421–434.

Rowe, H. (1998) Legal implications of consumer-orientated electronic commerce. *Computer Law & Security Report*, **14**(4), 232–242.

Schoop, M., Jertila, A. and List, T. (2003) Negoisst: A negotiation support system for electronic business-to-business negotiations in e-business. *Data & Knowledge Engineering*, **47**(3), 371–401.

Smith, A. and Clarke, R. (2000) Identification, authentication and anonymity in a legal context. *Computer Law & Security Report*, **16**(2), 95–100.

Stewart, R.A. and Mohamed, S. (2004) Evaluating web-based project information management in construction: capturing the long-term value creation process. *Automation in Construction*, **13**(4), 469–479.

Tah, J.H.M. (2004) Towards an agent-based construction supply network modelling and simulation platform. *Automation in Construction*, In Press.

Thelen Reid & Priest LLP (1997) *Commercial Users of the Internet Can Find Themselves Defending Lawsuits in Unexpected Places*. Available at http://www.constructionweblinks.com/Resources/Industry_Reports_Newsletters/Aug_28_1997/aug_28_1997.html (accessed 28 September 2004).

Thelen Reid & Priest LLP (1999a) *California is First State in Nation to Adopt Electronic Contracting Law*. Available at http://www.constructionweblinks.com/Resources/Industry_Reports_Newsletters/Nov_18_1999/nov_18_1999.html (accessed 05 October 2004).

Thelen Reid & Priest LLP (1999b) *Current Legal Issues Facing Businesses on the Internet: legal Risks and How they Can Be Avoided*. Available at http://www.constructionweblinks.com/Resources/Industry_Reports_Newsletters/July_1999/july_1999.html (accessed 05 October 2004).

13 Knowledge Management for Improved Construction e-Business Performance

Charles O. Egbu

13.1 Introduction

In many industries, the emergence of e-business has transformed organizations and industries. e-Business relies on the development of new business strategies based on existing and new networks. The digital connectivity allows multiple supply chains to form supply networks that have significant potential to improve the business value for stakeholders within the supply networks. The application of information and communication technology (ICT) tools is becoming increasingly important to enable organizations to carry out their business processes online. This provides a significant challenge for construction organizations to rethink their existing business strategies. It also provides opportunities for exploiting technology effectively to improve business performance. The construction industry is still in the throes of embracing e-business initiatives to effectively revolutionize some of the processes which it employs in developing product concepts, design and the delivery of projects to meet clients' requirements. e-Business initiatives provide opportunities for multiple partners to integrate their organizational processes via virtual networks. This provides a common platform for which inter-organizational knowledge can be harnessed and combined to produce a synergy that impacts upon product and service delivery. A knowledge-based business (k-business) is a knowledge-intensive one, where organizations use their intellectual capital (IC) to increasingly find new ways to add value to their existing business processes. It involves a creative and innovative combination of existing knowledge stocks with newly acquired, or created, knowledge to maintain a competitive advantage. An important strategy in exploiting the role of knowledge in e-business initiatives involves, *inter alia*, an organization knowing if it is ready to launch a knowledge-based business through the assessment of its knowledge assets via proper knowledge management procedures. The accurate knowledge of the availability of its knowledge 'stock' and proper cultural and technological infrastructure is vital. This chapter documents the key contributions which knowledge management can make to effective e-business initiatives and the factors that promote and inhibit the contributions of knowledge management to e-business initiatives. It also proposes and documents the key strategies and processes needed to fully exploit the benefits of knowledge management. In reflecting on the synergy between e-business and knowledge management the chapter also argues that however great the information capabilities of computers may be, it remains the case that information generated by computer systems is not a very rich carrier of human interpretation for potential action. Knowledge resides in the user's subjective context of action based on that information. To some extent, it is the understanding of this important point that paves the way to general understanding and fuller exploitation of knowledge for improved e-business initiatives in the construction industry.

13.2 Knowledge management in context

Defining knowledge management (KM) precisely can be problematic. Nowhere in the literature on KM is there a single unified meaning of the concept. Many authors have attempted to explain certain elements of KM, specific to their own academic domains. Much of the ambiguity associated with KM is rooted in these authors' epistemological beliefs. In order to understand KM, it is first necessary to understand what is meant by knowledge. The debate about the meaning of knowledge is a pastiche of abstract ideas, which is too substantial to approach in this chapter. However, it is useful to briefly document the meaning of knowledge and knowledge management in the context of this chapter. Knowledge management is about the processes by which knowledge is created, acquired, communicated, shared, applied, and effectively utilized and managed, in order to meet existing and emerging needs, to identify and exploit existing and acquired knowledge assets. Knowledge consists of truths, beliefs, perspectives, concepts, judgements, expectations, methodologies and know-how, and exists in different forms.

It cannot be ignored that knowledge management plays a vital role in helping to identify deficiencies in organizational skills and competencies (Egbu *et al.*, 2001). Knowledge management can help to address issues of culture and organizational structure that are seen to be possible obstacles to effective adoption and implementation of e-business processes in construction organizations. The adoption of new working processes like e-procurement systems will always face internal resistance by the workers. Furthermore, if employees do not understand the benefits associated with automating procurement processes there will be a lack of confidence and security in deploying this new technology. This, in turn, will cause delays and even reluctance to use the e-procurement system. Knowledge management processes help to identify these softer non-technical concerns and can help to design training and awareness programmes for the organization. Such an approach also helps to encourage the development of a more conducive environment for learning for organizational employees in meeting the needs of the organization. Knowledge management can facilitate the creation of an environment where employees are committed and motivated to share knowledge to the benefit of the organization (Malhotra and Galletta, 2003).

Knowledge management can also play a vital role in identifying knowledge gaps in existing business processes and integrating any organizational processes, knowledge databases or repositories that can be utilized to improve these business processes to complement any e-business initiatives.

This knowledge gap identification process plays an important role in helping to map out the capability of the organization before any e-business initiatives are considered. It also can help to identify an appropriate technical infrastructure that complements the existing business needs, and identify emerging business needs that may face the organization with the introduction of e-business initiatives. Knowledge management can play a significant role in making sure there is a seamless integration of the people, and processes with the appropriate technical infrastructure for e-procurement initiatives. Without this integration and on-going support, it will be difficult to operate a successful e-business initiative.

13.3 Exploiting opportunities in the fast-changing environment of e-business: A knowledge management perspective?

With the advent of Internet technology, there is now more room for construction organizations to compete locally and globally, because the widespread electronic linking of individuals and organizations has created a new economic environment in which space, time and size are less limiting factors.

Knowledge management was defined by Egbu *et al.* (2001) and others as the 'process by which knowledge is created, captured, stored, shared and transferred, implemented, exploited and measured to meet the needs of an organization'. This definition has practical application because it encompasses the whole area of creation, capturing, and systematic management and sharing and accessibility of the knowledge where e-business comes into play.

e-Business involves any 'net' business activity involving telecommunication networks that transform internal and external relationships to create value and exploit market opportunities driven by new rules on the connected economy. The implementation of e-business essentially requires appropriate knowledge assets to be readily available with appropriately trained personnel, hardware and software in place. e-Commerce, which is a microcosm of e-business, is to do with conducting electronic transactions which is the buying and selling of goods and services online. This allows organizations to purchase products and services online from their supply chain partners.

Construction organizations embark on knowledge management issues in search of improved efficiency of their processes so as to maintain competitive advantage. This involves organizations learning to use the many talents and capabilities of their people and effectively inventing new processes to remain competitive. This requires the organization to have access to the right knowledge assets for deployment to produce a service or product from the right place, person and importantly at the time needed. However, due to the extreme fragmentation of the construction industry demand and supply side, much of the knowledge created during the project process is either lost or kept isolated and involves more time and costs to be accessed by the members of the supply chain.

e-Business initiatives facilitate the exploitation of knowledge assets and in this regard KM and e-business complement each other. KM allows for the creation of the cultural infrastructure, which involves the process of discovering, creating, sharing of knowledge and the continued development and use of knowledge. e-Business initiatives provide the channel for conducting business as part of an individual's (or groups of individuals) day-to-day work in decision making (Malhotra, 2000; Fahey *et al.*, 2001). Knowledge management in this context is intended to create, share and leverage increasingly higher quality knowledge facilitated by e-business in order to achieve three interrelated goals:

- To achieve superior external performance, including marketplace and financial returns.
- To achieve superior internal operating performance through operating efficiencies.
- To enhance the quality of life of individual members of the organization.

Some of the processes which KM and e-business can facilitate in the construction industry include the timely exchange of appropriate and accurate information/knowledge assets to the right person(s). This includes: tenders, enquiries, quotes, dispatch notes, invoices, credit notes, valuations and site instructions. For e-business initiatives to be effectively applied, integration must be implemented throughout the organization and with external business partners. It enables effective partnering, leading to reduced costs and increased benefits. Some of the tangible benefits that can be achieved by e-business in this way are:

- Tender distribution costs can be cut significantly. Tender documents can be placed on the Internet and contractors and subcontractors can price them and return them to the issuing authority. This can be achieved without the use of paper, avoiding delays in delivery. Tender entry time can be dramatically reduced.
- Reduction of invoice registration costs.
- Re-keying errors, delays and disputes can be avoided.

- Construction organizations can procure materials, plant and other resources at cheaper and better deals that may be necessary to perform their business outside from outside their traditional supply chain.

The proper application of e-business initiatives will complement KM and deliver improved efficiency to the organization. For that to happen the first thing any organization must do before implementing e-business is to carry out a thorough organizational performance analysis (using tools such as VERDICT, which is described in Chapter 3). This includes their strengths, weakness, opportunities and potential threats. This will help to identify their current position in relation to their knowledge assets and overall organizational capability. Some key measures that must be taken include:

- Provide real leadership for all knowledge management issues within the organization. This includes identifying the appropriate knowledge assets within the organization and designing the most appropriate strategy to exploit these knowledge assets through e-business.
- Ensure that the KM procedures in place are adequate and appropriate to the type of business.
- Institute current and on-going training programmes for appropriate personnel that will implement the key KM and e-business initiatives.
- Integration of the organization with other organizations along the supply chain for sharing and acquisition of new and relevant knowledge.
- Upgrade the knowledge assets in the organization to meet the demands of the clients/customers. Learn from other successful organizations and choose appropriate e-business technologies. This includes both hardware and software. This could be achieved by using outsourcing alliances as a business model (Kalakota, 2001).

The above measures will help organizations mobilize their knowledge assets in readiness for deployment. This can only be achieved if there is a visionary leader driving the initiative.

13.4 Organizational challenges in using the internet to commercialize knowledge assets

As e-business initiatives increasingly offer the platform for new forms of marketplace strategy models, organizations themselves are faced with a host of challenges in using the internet to commercialize their knowledge assets. These are mainly human, cultural, social, organizational, legal and technical, and include:

- The change of organizational culture to embrace e-business as a core element of business strategy.
- Constantly/rapidly changing organizational culture of business partners (clients, customers) to cope with changing market conditions.

- The inability to cope with the fast-changing tools that are offered on the market. This includes the cost of purchasing and training to use the tools.
- Management task to align business strategies, processes, and applications quickly, correctly, and all at once.
- Agility is the key to survival in the new economy; but construction organizations are slow to respond to changes.
- Lack of strong and decisive business leadership, and inability to spot trends quickly and create effective business strategies.
- The lack of technical expertise available within organizations, especially small and medium enterprises.
- The cost of acquiring Internet-based software and hardware tools to facilitate their business processes.
- The uncertainty and risks of changing from the 'old' business model to the 'new' business model. This is more so for an industry, which is already a high-risk industry and very conservative – not willing to take further risks.
- The legal issues associated with contractual liabilities of online procurement of goods and services.
- Security of online procurement.
- Technology shifting power to the buyers (consumers) as e-business is changing the channels through which consumers and businesses have traditionally bought and sold goods.

Few would argue about the importance of organizational culture as an important facet in embracing e-business and the challenges this poses to organizations. Theories about organizational culture favour the evolution of a 'community of practice' where social interaction of employees leads to a knowledge sharing culture based on shared interests, thus encouraging idea generation and innovation. In all organizations, the politics of knowledge sharing is an issue. Employees and employers from diverse backgrounds often come into conflict over important decisions. Manipulating these tensions to achieve 'creative abrasion' is a strategy to maximize innovation. However, it is a challenging task that involves disciplined management. Of course, leadership is an inherent part of organizational culture, but also extends into areas of strategy and structure. Leadership is an organizational responsibility which is of great value in creating the structures, strategies and systems that facilitate innovation and organizational learning. Leadership can also help to build commitment and excitement, collective energy and empowerment. Moreover, empowering employees to generate and share knowledge is the task of management. For example, implementation of rewards and punishment schemes can be a stimulus for successful KM. Motivating employees to share the tacit knowledge which they have involves good people management, where trust is itself an incentive.

The above discourse would suggest that however great the information capabilities of computers may be, it remains the case that the information generated by computer systems is not a very rich carrier of human interpretation for potential action. Knowledge resides in the user's subjective context of action based on information. The understanding of this important point is vital to an understanding and fuller exploitation of knowledge for improved e-business.

The challenges above highlight the softer issues of management that must be adequately addressed before any e-business strategy can be successfully implemented. In order to overcome these challenges facing organizations in exploiting the opportunities presented by the Internet, there must be a clear vision for change from top management. This must be backed with evidence of support. Secondly, the organization must define its business processes and develop a 'business model' that suits the organizational needs for implementing the new method of doing business via the Internet. Thirdly, there must be appropriate awareness and training provided for the employees of the organization. Should the organization lack adequately trained personnel to implement the e-business strategy, it can employ skilled personnel as part of the organization or engage consultants to provide this service. Alternatively, the organization may source technological expertise and tools by forming alliances with members of the supply chain who may already have these systems operating in their organization. In this way the organization can benefit from the successes of 'an integrated system'. This may help to overcome the time lag taken to reach the learning curve, thus helping the organization to make the transition within as 'short' a time as possible.

The need for training stands out as the key area for organizations to pay attention to. Training will help broaden the horizon of the organizational employees to see opportunities for innovation in the new business model. Training also prepares the employees to shift their mental models from the old operating procedures to the new operating procedures. It also allows employees and management to identify gaps of knowledge needs and helps to devise ways to bridge the gap in knowledge.

Without proper KM in place it will be difficult for any organization to effectively implement e-business strategies in a sustainable manner. In a changing marketplace, an organization's success invariably depends on the ability to innovate and integrate new technologies into service offerings. It will be practically impossible to be innovative without proper KM implementation strategies. Knowledge management will help executives become proficient trend spotters when implementing e-business initiatives; if they do not, their companies are likely to do badly. The next generation of e-business organizations in the construction industry must be imaginative in order to radically change the value proposition within and across the industry.

13.5 Knowledge assets employed by construction organizations in e-business initiatives

In this 'knowledge economy', organizations are depending more on their knowledge assets or IC to remain competitive. Knowledge assets are seen as the key currency required for trading online by organizations intending to pursue e-business. In the main, knowledge can be said to exist in two forms, tacit and explicit. Tacit knowledge is that which is held in peoples' heads and is very difficult to capture or transfer. Explicit knowledge is knowledge held in documents, reports, databases and patents. While it is recognized that knowledge is key to implementing e-business, it is important that there are appropriate processes, techniques and technologies in place within organizations to fully identify the appropriate types of knowledge, in its various forms, for their successful exploitation via the Internet when conducting e-business. Table 13.1 presents a list of the key organizational knowledge assets used in e-business in construction.

Tacit knowledge is generally seen to contribute the greatest to competitive advantage, as it is difficult to mimic. The capturing of the tacit knowledge assets that reside in staff and members of the supply chain poses a great deal of challenge for organizations. Knowledge can also change form when it is passed from one source to another. The context of its creation

Table 13.1 Key knowledge assets employed in e-business initiatives

Knowledge embedded in organizational staff	Tacit
Electronic databases	Explicit
Working reports and other publications held in electronic format	Explicit
Standardised organizational forms in electronic format (e.g. Health and safety audit forms)	Explicit
The collective knowledge of members of the supply chain/partners – accessed or shared face-to-face or via electronic networks	Tacit and explicit
Knowledge of clients/customers of organizations	Tacit and explicit
Organizational knowledge embedded in business and operational processes (including in manuals/brochures)	Explicit
Knowledge embedded in software programs	Tacit and explicit
Specialist knowledge in the engagement of specialist third parties (e.g. in outsourcing and in the hosting of Websites and e-tendering portals)	Tacit and explicit
Organizational memory (the total organizational knowledge explicit and tacit)	Explicit, Implicit and Tacit
Patents held by the organization	Explicit

and application also changes depending on the initial context in which the knowledge was created or captured, stored and the intended need which it is expected to meet. This dynamics of knowledge assets call for its constant updating so as to maintain its currency and relevance for changing customer needs and market conditions.

Explicit knowledge is also an important asset. Knowledge captured and made explicit needs to be converted into formats that can be easily accessible by various e-business processes as and when needed. This requires knowledge to be held in electronic systems and formats where organizational staff and/or partner organizations in the supply chain can have access as and when needed.

The level of usage of different types of knowledge assets is a reflection of the extent to which e-business initiatives are implemented in the organization and the level of resources allocated and support from the top management. It is also likely to change as demand for certain knowledge types changes.

13.6 Organizational readiness to lunch a knowledge-business (k-business)

By appropriately combining the potential of the Internet with an explicit approach to commercialize its knowledge assets through e-business, construction organizations can create an effective and competent knowledge business (k-business). A knowledge-based organization involves a creative and innovative combination of existing knowledge stocks with newly acquired or created knowledge to maintain competitive advantage. An organization will know if it is ready to launch a k-business through the assessment of its knowledge assets and proper KM procedures. An accurate knowledge of the availability of its knowledge 'stock' and proper cultural and technological infrastructure implemented will ensure that an organization is ready to launch a k-business.

There is currently little or no formal tool or methods available to help construction organizations measure their capability and competencies to launch into e-business which formally incorporates knowledge management principles. Vines and Egbu (2004) have earlier argued for one. Such an e-business readiness assessment protocol (e-BRAP: Vines and Egbu, 2004) could be operationalized within a three-phase approach that is the primary phase (organizational environment), the secondary phase (the organizational knowledge and technical infrastructure) and the tertiary phase (external stockholders identification and integration).

The variables in each phase (Table 13.2) of the e-business readiness assessment protocol can be measured by their readiness rating scale from 1 – 'not conducive at all' to 5 – 'very conducive'. Readiness rating scales from 1 to 3 will not qualify the organization to proceed onto the next phase because

Table 13.2 Key issues within a three-phase e-business readiness assessment process protocol incorporating knowledge management principles

Phase	Variables/key issues	Processes
Primary phase (Organizational environment)	Examine the opportunities and threats facing the organization and industry.	Top management vision, commitment and leadership; employees' buy-in; resource allocation.
	Examine existing strategy and include e-business as an important aspect of the business.	
	Adjust organizational culture to embrace e-business concepts.	
	Adjust organizational structure (reduce layers to make flatter) to respond to internal and external business needs effectively.	
Secondary phase (Organizational knowledge and technical infrastructure)	Design and implement organizational Knowledge Management strategy.	KM must be integral part of business process.
	Map out the key organizational knowledge assets important to support e-business	Education and training programmes must be designed and implemented.
	Institute education and training strategy to sustain e-business.	
Tertiary phase (External stockholder's identification and integration)	Implement strategy to integrate supply chain onto a common e-business platform/infrastructure.	Buy-in support and commitment from supply chain.
		Engage supply chain to invest in integration process.

even if the organization embarks on an e-business initiative with variable ratings in this range, the chances of increasing return on investment may be significantly affected. On the other hand, a variable that has a readiness rating level at 4 or 5 suggests readiness to implement e-business and there is therefore a high chance that the organization can see success in its e-business strategy. The variables presented here in the proposed 'three phase' process protocol would be key variables. Organizations can include other variables or sub-variables which may be particularly important for their organizations at each phase of the process and apply the same readiness assessment rating.

At the primary phase, when considering implementing e-business, it is important that an organization investigates the market it is operating in and the opportunities and threats confronting it, for example, due to the impact of technological advancement. The level of opportunities and threat facing the organization will determine whether e-business can be implemented to overcome the threats and exploit these opportunities. Basically, it is a position audit of an organization within its market of operation. A very simple rating of conduciveness for each key variable which impacts on e-business is useful to determine the organization's level of readiness and suitability to engage in e-business. These variables include organizational culture, structure and strategy conducive for e-business. All factors in each phase should score a rating of 4 or above to enable an organization to progress to the next phase. This assessment can be done in parallel with other phases and is an iterative process. The level of judgement is based on each organization's thorough evaluation of where it stands with regard to the identified variables when considering an e-business initiative.

For any e-business launch to remain successful there must be a knowledge management initiative in place. The implementation of KM is extremely important to supply the appropriately 'packaged' organizational knowledge to meet customer needs when deployed via e-business initiatives. This is the key focus of the second phase. It is vital that the KM assessment rating is 4 or above which will indicate its ability to sustain e-business after the launch. However if the organization does not have appropriate personnel, it will be impractical to effect e-business to successfully exploit the knowledge base of the organization. Therefore, it is vital that education and training programmes are designed and put in operation to continually renew the organizational knowledge base to meet fluctuating customer requirements when implementing e-business initiatives.

At the tertiary phase, an organization has successfully created a business strategy to implement e-business and the organizational culture and structure is conducive to implement e-business. This internal organizational conduciveness will then provide the basis to integrate the external supply chain. The integration of the supply chain is key to unleashing the full benefits of implementing e-business. At this phase the supply chain is operating as a single entity and is capable of meeting demands as and when required. The integration enables the supply chain knowledge to be exchanged to produce a synergy to meet business demands.

Several criteria worthy of consideration within such an assessment regime include:

- If the assessment of the success of e-business strategy is to be related to their contribution to business objectives, the assessment protocol should embrace all the key organizational elements: IT infrastructure, people, business processes, culture and the work environment;
- The e-business protocol should be able to provide a general assessment of the 'current status' of an organization (i.e. the current state of readiness).

- The protocol should be able to assess the 'required status' of an organization for a particular e-business initiative (i.e. the required state of readiness).
- The protocol should adopt the maturity-level techniques to facilitate the measurement of the 'Readiness Gap' that is the gap between the current and the required state of readiness, prior to the implementation of an e-business initiative. It should also identify the necessary guidelines for an organization to progress through the maturity levels.
- Each maturity level should provide guidelines for managers to improve the readiness status of their organizations through appropriate awareness, education and training.
- The protocol should be of a holistic nature to provide organizations with a quick and general assessment regarding their readiness gap.

13.7 Conclusions and recommendations

The construction industry is a knowledge-intensive industry. However, due to its level of fragmentation, some key knowledge assets are 'locked-up' in organizational and divisional 'silos'. This creates a situation where organizations spend more resources and incur additional costs in trying to meet a customer's need by trying to locate or create already held knowledge. Knowledge management enables construction organizations to identify key knowledge deficiencies.

e-Business requires organizations to refocus and reconfigure every type of tangible and intangible assets including the many forms in which knowledge is held. Knowledge management also provides that reconfiguration and refocusing of the key knowledge needed to launch, drive and sustain e-business.

e-Business initiatives increasingly offer the platform for new forms of marketplace strategy models for construction organizations. But organizations themselves are faced with a host of challenges in using the Internet to commercialize their knowledge assets. These are mainly human, cultural, social, organizational, legal and technical. Knowledge resides in the user's subjective context of action based on information. To some extent, it is the understanding of this important point that paves the way to an understanding and fuller exploitation of knowledge for improved e-business initiatives in the construction industry.

References

Egbu, C., Botterill, K. and Bates, M. (2001) The influence of knowledge management and intellectual capital on organisational innovations. In: *Proceedings of the 17th Annual Conference of the Association of Researchers in Construction Management (ARCOM)*, University of Salford, 5–7 September, 2, pp. 547–556.

Fahey, L., Srivastava, R., Sharon, J.S. and Smith, D.E. (2001) Linking e-business and operating processes: The role of knowledge management. *IBM Systems Journal*, **40**(4), 889–907.

Kalakota, R.M.R. (2001) *E-Business 2.0: Roadmap for Success*. Addison-Wesley, New Jersey.

Malhotra, Y. (2000) Knowledge management for e-business performance: Advancing information strategy to 'Internet Time'. *Information Strategy: The Executive's Journal*, **16**(4), 5–16.

Malhotra, Y. and Galletta, D.F. (2003) Role of commitment and motivation in knowledge management systems implementation: Theory, conceptualization, and measurement of antecedents of success. In: *Proceedings of the Thirty-Sixth Annual Hawaii International Conference on Systems Sciences. Los Alamitos* (R.H. Sprague Jr. ed), CA: IEEE Computer Society Press. Available at http://csdl2.computer.org/comp/proceedings/hicss/2003/1874/04/187440115a.pdf.

Vines, M. and Egbu, C. (2004) Readiness assessment process protocol for e-business initiatives in construction organisations. In: *Proceedings of the 20th Annual Conference of the Association of Researchers in Construction Management (ARCOM)* (F. Khosrowshahi ed). Heriot Watt University, 1–3 September (Vol. 1) ISBN: 0953416194, pp. 595–602.

14 e-Commerce in Construction: Industrial Case Study

Tim C. Cole

14.1 Introduction

The regular exchange of commercial data between construction industry supply chain partners, though a marginal participation sport during the 1990s, has become an increasingly regular sighting in the new Millennium. This chapter looks at how this evolved and reviews the practical experiences of early adopters to help foster still wider adoption. This also provides a guide to companies as they seek to achieve maximum gains from their efforts to implement e-commerce.

14.2 Background

One of the primary objectives of e-commerce is to integrate businesses at a technology level without constraining innovation or process flexibility. One of the reasons the 'dot com' era resulted in so many false dawns was because the focus became skewed towards pure technology to the detriment of decades of business experience. Technology and commerce are now working in closer partnership, delivering beneficial and lasting change.

In this day and age technology underpins nearly every aspect of commercial operations. This makes business process integration between computer systems an obvious objective. When business was managed and transacted on paper, then paper was the logical means of communication. As accounting, procurement and other processes became reliant on software rather than paper; it was increasingly a folly to retain paper as the means of communication. For transaction data, such as invoices and orders, the use of paper or even unstructured files (e.g. Word Processed documents or email text), makes little sense, as they cannot be imported into a receiving application without manual intervention. Manual data transfer is a clear hindrance to further process improvement and is costing businesses time and money.

During the 1990s the Internet and email enabled faster communication and some integration. At that time though, very few back office systems were able to enjoy meaningful dialogue with any other back office systems.

Having embedded process responsibilities within computer systems, the next step was to understand the complexities of application-to-application communications without printing and/or re-keying information. Failure to integrate at the transaction level would result in efficiency gains being constrained by the very technology that had opened so many new horizons.

14.3 A historic perspective

History has shown that the progress of innovation is an incremental one. New ideas build on top of the foundations laid down by previous innovators. So it is that, in achieving operational e-commerce, there have been a series of small steps taken over recent years. Each step has brought companies closer to the goal – including those steps that achieved little more than identifying dead-ends.

14.3.1 Common data standard

One of the early stages in the development of what is now known as operational e-commerce was a drive for a common data standard that would become the VHS (Video Home System) equivalent for electronic communication in construction. There was wide support for the concept, but bringing this concept to reality would be a complex and arduous process. At the time there was a wide range of data formats already available. Just as Martin Luther King 'had a dream', many others, including the author, shared a dream. In this case, it was that all computer systems would learn to export or import documents using just such a single data language. This would allow computers to communicate without the problems us humans experience in handling the many languages spoken around the world. Several process integration experts devoted many man-years to develop data standards for use in the construction industry. These were mostly based on the international UN/EDIFACT (Electronic Data Interchange for Administration Commerce and Transport) protocols, which were compiled to flexibly support processes across national boundaries and industry sectors – ideal for the gregarious nature of our industry. The resulting standards, mostly developed under the CITE (Construction Industry Trading Electronically) initiative, provided a highly competent vehicle for commercial data exchange. Embedding the standard within different applications and getting companies to commit to their operational use, however, proved harder to achieve than it had been to gain support for the initial objective. Competing development priorities and the failure to convert user interest into commercial pressure were just two contributing factors.

A majority of relevant software, during the mid to late 1990s, was either unable to exchange electronic documents or did so using a proprietary format. This resulted in both the sender and receiver having to install a data translator to convert to/from any standard language. These were not insurmountable barriers, but were enough to prevent adoption at any meaningful level.

14.3.2 The role of the Internet

The explosion of Internet access has opened several opportunities for electronic communication that were previously not possible. The complete over-reaction to the impact of the Internet brought, at the very least, a false dawn and certainly caused a review of implementation plans by many companies that delayed operational benefits by a number of years. The balancing point is that, due to the Internet, companies can now move forward and adopt e-commerce at lower cost. The overall impact of the Internet is clearly a positive one.

To many, the arrival of the Internet's structured data language, XML (eXtensible Markup Language), offered the ultimate solution for computer-to-computer communications. In reality, however, XML is nothing more than another data formatting protocol. It is an Internet-friendly language but, on its own, would not make data exchange an overnight success. Even with XML, users still are required to define how the data elements within the message, such as an invoice date, item description, item quantity or price are described. The fact that this work had already been undertaken by initiatives such as CITE, made little difference. During this period, new standards were continually being developed by almost anyone who had the time and inclination. According to James Whittle, of eCentre[UK] (Whittle, 2002), there were over 2000 XML versions of an invoice in place before entering the current Millennium. This was no basis for a standardized approach to data exchange. This predicament meant that there was neither widespread adoption through the Internet nor a common standard upon which to build an integrated future.

14.3.3 From common standard to a hub-centric environment

A new approach was required that would harness the benefits of the Internet and the strength embedded in the international data standard. Yet, even for this to succeed it would be essential to find a way to integrate applications without requiring that they themselves 'had' to have adopted any data exchange standard. Delivering against these three components was felt to be the best way to remove the inertia that had constrained earlier initiatives.

Developments in this field have meant that today it is rare to find an application that cannot export or import electronic documents. The format is often unique to the specific application, but is normally able to convey the level of business data needed to facilitate e-commerce. The next step on the innovation trail was to develop services that could

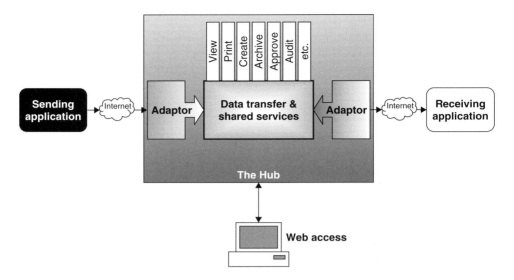

Figure 14.1 Basic hub elements

import or export data seamlessly from all the different applications in use. Also, communications could harness the capability of the Internet and the central-hosted services could facilitate the business process logic required to truly replace paper. These services are normally referred to as 'Hubs' (Figure 14.1), though there is no standard definition to set out the range of services that have to be provided in order to be known as a Hub. Data integration can either be handled by the Hub or each trading partner may decide to conform to a pre-existing interface readily supported by the service provider. Experience has already shown the value of establishing an interface with maximum data content at the outset, rather than using a limited data set, even though such an option may seem expedient in the short term. Like most technologies, e-commerce requires scaling and a weak interface will soon be challenged by the needs of a company's trading partners.

There are cost and time benefits through the use of such Hub services, but this is not to say that Hubs are all the same. As more companies join Hubs to facilitate the exchange of electronic documents there should be pressure to raise the minimum service level to cover the range of process issues covered in Section 6 in this chapter. Hubs offer a way towards the ultimate goal of full integration at the application level, through their ability to deliver scalable connections, business process support and shared services. The establishment of a 'Hub Alliance' (www.huballiance.org) has further helped in this regard by providing for the open interconnections between different Hub providers. As with the links between different mobile phone networks, the Hub Alliance expands the accessible community and so helps to accelerate adoption through enhanced value available to companies looking to join any Hub service that is a member of the Alliance. The central control that Hubs provide helps companies to connect earlier than was previously possible. Internal IT resources are often a scarce commodity and setting up multiple back office links with trading partners would be hard to justify and certainly creates a growing maintenance burden upon these resources.

Hub services need to facilitate a coordinated approach to e-commerce implementation between the document sender, receiver and their respective applications. This is about much more than moving data from Company A to Company B. Users of Hub services require a scalable solution that can underpin e-commerce growth and which delivers essential functionality, including:

- Integration capability
    o Able to readily link to different applications (e.g. accounts system)
- Interconnect policy
    o Operates an open interface with other commercial services (e.g. Hub Alliance)
- Ability to support business rules
    o Allows business processes to be replicated within the electronic environment (Sender and receiver)
    o Supports the EU Invoice Exchange directives
- Support for sufficient document types (invoices, orders, etc.)
- Security
    o Dual hosting, encryption, disaster recovery, independent service testing
- Access to shared services
    o Progress monitoring, document approval, workflow, status updates, etc.
- Support full archiving and application outputs
    o To ensure the ability to remove paper completely.

14.4 e-Commerce implementation: Practical issues and benefits

When electronic commerce is to be implemented as an integral part of day-to-day operations, there will inevitably be a powerful, sometimes fractious, coupling between commercial reality and technology. The result of such a marriage can be either a disappointing shock to each side, or a synthesis that delivers real opportunities. Business managers understand their operational responsibilities and rightly demand highly flexible and technically competent solutions that have a similar depth of process awareness. It is therefore essential, in developing electronic business services that the development process engages the industry needs and the available technology within an informed, innovative and collaborative undertaking.

There are hundreds of thousands of trading relationships across the UK construction industry with many companies having individual supply chains running to thousands of companies. At the same time, there are numerous procurement, cost management and accounting applications in use, most of which operate largely on a stand-alone basis. These complex trading relations, together with disparate applications, pose a significant challenge to any new (technological) implementation. When looking for the benefits of e-commerce implementation it is important to keep the short- and long-term in mind. There are still those who count reduction in stamps consumed as a value marker, but the goal is much bigger – and one that will see winners and losers. Cutting the cost from wasteful manual processing is indeed the level one saving. However, even at this level, the value of reduced queries, faster processing, streamlined archiving and improved supply chain relationships should be counted. At level two, process reengineering can unlock costs and time constraints throughout the business. Finally, at level three potentially the greatest rewards result from improved process visibility and control. For example, commercial negotiations can trade price discounts for an improvement in payment terms. For such goals to be achieved they have to stand on an open, consistent and trusted transactional relationship. Some companies have already reported that they are winning more work through the added value of the e-commerce enabled process efficiencies they can offer to their customers. Given that the operational experience of early adopters has already reported savings from £2 to £7 per document, the value proposition is rarely a barrier to progress. The challenge is agreeing the steps to take to unlock the savings.

Effective implementation of e-commerce improves the efficiency of, or facilitates enhancements to, business processes. Change must centre on the needs of the business. It is business managers who understand the operational and process bottlenecks that drain costs and who should rightly be fully engaged in planning the adoption of e-commerce services. Solutions must match the flexibility and technical competence that underpins the commercial engine driving each successful company. The technology must fit to the required process and protect operational competence without compromising innovation or business growth.

The procurement process may comprise generic elements, such as orders, delivery notes, invoices and payment, but is multifarious in its deployment within different companies. Computer applications vary; processes for generating documents vary; incoming document management will vary; authorization rules vary; reconciliation rules vary; payment processes vary and so on. There is no right or wrong, which means that e-commerce implementation must be sufficiently flexible and competent to meet the business process demands of each company.

Initially pragmatism will require more effort to achieve integration between different back office systems than companies might like, but real gains can be made right now.

14.5 The first adopters

There is no getting away from the lubricant effect of mass action. Although the benefits had almost universally been accepted, while the bulk of construction companies remained in paper-land, there was little imperative to take on the inevitable change process concomitant with the adoption of e-commerce. Once it became clear that the first companies to adopt e-commerce were reporting tangible benefits, competitive forces and a clearer business case justification helped to increase the take-up rate and market perception considerably.

In practical terms, this breakthrough was achieved by pooling the commercial impetus of a number of main contractors and suppliers. Collectively, the value proposition was compelling and the challenge was then to marry this with a technology solution able to facilitate open exchange of data. In turn, the success of the technology required the support of the contractors and suppliers. The visionary support from the early adopters ensured that construction companies established a basis for future e-commerce in construction. The later part of this chapter reviews some of the practical experiences of the first adopters.

The deployment of technology within the construction supply chain has always raised fundamental issues that, if not addressed, can turn opportunity into disappointment. There used to be a 'hot potato' debate within companies as to who would be responsible for assessing and implementing electronic data exchange activities. Practical experience shows that there are three key representatives required to support a successful e-commerce implementation.

- Business Sponsor, normally a senior executive;
- IT Support, able to manage the solution integration; and
- User Representation, normally the person who will be responsible for the users of the services and its operational impact.

Although not necessarily three different people, without these three elements being engaged in the e-commerce implementation there is a clear risk that the process could run late, incur additional costs or even fail. Further, once the implementation is complete, the take-up rate could be slower that anticipated. It perhaps goes without saying that gaining support from all three areas should be driven by the business sponsor.

e-Commerce services must retain the ability for each company to innovate at a process level. e-Commerce is not about neutering all trading partners at the information level, as this is precisely where innovation can deliver differential benefits. Integrated data services must be flexible at the deployment level to allow each company to retain and enhance their own process and commercial value propositions.

14.6 Implementation issues: Case study examples

In this section the issues most often encountered during e-commerce implementations are tackled one by one. Most are drawn from situations arising with electronic data exchange hubs, but the general principles are applicable to all aspects of e-commerce.

14.6.1 Standard specification dilemma

It is a sad fact that true standards are rarer than one might like to think. Standards applied in a uniform manner so that the uses of the data fields by Company X are exactly the same as the use of the same fields by Company Y are sadly rare. Standard interfaces are often found to have been interpreted in a non-standard way. A simple example could be where a field requires the cost, but is not explicit whether this is before or after discounts, VAT, etc. Such anomalies require an element of customization to be applied.

Specification problems can apply to many different applications and so it is essential to 'dig behind the obvious' to check how best the application can interface to the world of e-commerce. Where a file specification exists for data export(s), it is strongly recommended that a sample file is created and cross-referenced against the specification at the earliest time. Test the values found in key fields, especially those with numerical content and always dig deeper than the syntax. For example, check whether the price rate is correctly showing as 'Net' or 'Gross' and if the overall cost include discounts or not?

Where receiving data, the checking process is a little harder as users cannot produce a test file by themselves. Where concerns exist, users should try to get a small test file created in the predicted format and test that this loads successfully.

There is little worse that building an interface to an e-commerce service only to find on the day testing begins that your beloved application is singing off quite a different hymn sheet to that you predicted.

14.6.2 The commitment issue

Although this point has been covered earlier in this chapter, the importance of commitment cannot be emphasized enough. It can prove to be one of the most frustrating problems and certainly a potential cause of serious delays.

It is sadly not unheard of for a company to complete a connection to an e-commerce service and then wait months, even years before it is put to commercial use. There must be a clear business case backed up with clear ownership of the implementation by senior management.

14.6.3 Static data

Static data is data that changes very rarely, if at all. A good example would be the address of a company or perhaps their VAT number. The relevance to e-commerce of such information is that it was normally pre-printed on stationery and is often not present within data files exported from back office applications. This can impact on a company's ability to readily provide the full equivalent of a paper document in an electronic form and could even lead to a compliance problem with the minimum content for an invoice as specified under the EU Directive on electronic invoicing (EU Directive, 2001).

The ability for a service to enhance the data feed, by adding static data, is therefore very useful. For those about to change their back office applications or create a data interface there is a chance to resolve this issue permanently. It is recommended that a company's data interface contains the same level as detail as the equivalent paper document or perhaps even more. This is especially important given that applications are good at ignoring unwanted data, but lousy at creating data that is not there!

14.6.4 Document referencing

Because a company's back office system (e.g. their accounts application) is managed in isolation, there is rarely a common form of referencing between two such systems. For example, to 'buyer A', 'supplier B' will be known by a unique account reference, while 'buyer A' is known to 'supplier B' by a quite different account reference and it is unlikely that either system knows how this is handled by the other. Another example is in deciding whose product identification system should be used, buyer A's or supplier B's, the manufacturer's, or a combination of the three? This predicament can make electronic orders less attractive to suppliers especially if they are forced to intervene manually to ensure the right product is identified and supplied.

Once again, standardization would not go amiss, but it is essential that this is planned during the implementation phase. Hubs can use look-up tables to insert the correct account numbers to ensure the receiver can process the data automatically. Being able to resolve such an issue can determine whether documents can be automatically loaded into a receiving application or whether each will still require a level of manual processing.

The good news is that, as more data is exchanged electronically, the opportunities to resolve this problem increase. For example, incoming orders could automatically cross reference a pre-existing catalogue or an incoming invoice could cross reference an order. Hubs can use the outgoing data to augment any incoming data to improve automated import processing.

14.6.5 Archiving

There is clearly less to be gained from sending electronic documents if you still have to send the paper copy as well. The recipient needs sufficient data to allow the paper copy to be made redundant. This is one of the reasons why the importance of exporting as much data as possible has been stressed in this chapter. There will, however, always be two targets to hit when sending electronic documents. The first target to hit is the receiving application and the second target is the management information or archiving system. Clearly one file could be used to achieve both, but two files are very often required. The application file rarely contains the full data-set, but the ability to resolve document queries demands a paper equivalent. It is for this reason that both scenarios need to be provided. Hubs therefore may have to provide two outputs with the archive version containing all available data. The full data version can be stored offline and fed into an archive system to remove the need for paper scanning whilst retaining a paper-equivalent representation. With the increasing use of Document Management systems, the ability to satisfy the application and operational views of trading documents is increasingly of interest to companies.

There is a clear upside from using XML when managing application and user views of electronic documents. This is because such files can readily be interpreted by software whilst at the same time supporting visual presentation using what are known as 'style sheets'. A further benefit is that companies can lay out all documents received electronically on the screen using the same template, so improving the speed at which operators can assimilate information when required.

14.6.6 Business rules

The addition of static data and the use of look-up tables, both form part of this general topic. Receivers of documents are unlikely to want to simply dump data into their application. It is, however necessary to encourage a move away from re-keying data, bearing in mind that one of the most flexible computers available is involved in this process – the human brain. In general people tend to apply rules to the data they extract from paper documents. As mentioned already, they determine the correct account details, identify the correct product or check that key data is present. All these decisions are what are referred to here as 'Business Rules' and must be accounted for in any automated data processing.

Both senders and receivers must be able to define and maintain business rules within the exchange process. This normally means within the

Hub service or within a modified local interface that allows rules to be imposed before final loading of the data to the back office system.

One of the most obvious advantages that such a process capability can facilitate is the trapping of documents that will clearly fail to process correctly were they to be loaded. For example, invoices that do not contain an order number (or perhaps through the use of a look-up table – a valid order number) can be rejected and returned to the sender. Operating experience has shown that document senders, especially suppliers submitting invoices, welcome the buyer imposing such a rule within a Hub service. Rather than waiting until they are chasing payment to discover there is a problem they will find out straight away. Both parties avoid the cost and time wasted chasing up errors that can readily be fixed at source.

An e-commerce solution without the ability to support business rules will have limited scope as a facilitator for open trading.

14.7 Specific case study examples

The following sections are based on real case studies, while protecting the names of the parties concerned. In each case the details are limited to the solution adopted so as to focus on showing how e-commerce implementations have addressed real business scenarios.

14.7.1 Buyer A

Buyer A, a medium-sized contractor, was keen to start exchanging electronic documents to/from their back office system but when looking to take the first step realized that they could not register invoices unless delivery data had previously been loaded and approved. The solution was to upload a simple look-up table from the back office system to the Hub containing details of the approved deliveries which could be used to cross-reference incoming invoices. Where a match can be found, invoices are downloaded. Where no match is found, invoices are held in a pending process. Subsequent uploads of look-up data could be used to release more invoices and, in exceptional cases, manual intervention could be used to ensure that no invoice ever remained trapped in the system.

14.7.2 Buyer B

Buyer B, a major contractor, operated invoice approval through a separate application to their back office accounts system. Each system, however, had different data requirements, but required to be able to cross-reference from one system to the other. The solution was to use both the application and archive file options mentioned earlier, downloading both together.

Each of these two files adopts the same filename, but with a different file extension to facilitate the cross referencing.

14.7.3 Supplier/Buyer C

This company, a major construction material supplier, required the ability to trade with its customers and suppliers using several document types (e.g. invoices, orders, etc.) and had a number of pre-existing data exchange links in place. A number of electronic exchange interfaces had evolved over time as they responded to customer requests. These were becoming more difficult to manage as they increased their level of electronic trading. The solution adopted was to establish one strong interface between their back office system and the hub and then to replicate all the different interfaces as inputs to or outputs from the single hub. The benefit was a much simpler and more scalable process for the company while their trading partners were unaffected – other than that their trading partners had the additional benefit that they could expand their own electronic community through the access they now had to others connected to the hub.

14.7.4 Buyer D

Buyer D, a major contractor, had established a procurement process that linked documents such as Order Acknowledgements and Invoices with notifications sent to site for confirmation and cross matching within the back office to streamline processing. In this case the solution was to replicate the actions (such as the receipt of a data file) that triggered the established process actions. To ensure that the data complied with process requirements, business rules were used to reject invoices that did not contain key references.

14.7.5 Supplier E

This final example is the simplest to explain, but was one of the most effective e-commerce implementations with which the author was involved. Supplier E, an international material supplier, was a company used to working within the boundaries of strict project management procedures. From the outset, key personnel were identified and empowered to manage the overall project. The business sponsor liaised with their trading partners and a kick-off meeting held followed by regular teleconferences during the development and testing phases. Once complete, the project was signed-off and moved to roll out with their trading partners. The point is not just to promote the value of project management, but to underline the fact that e-commerce is a long-term venture that is worth building confidence across the supply chain by getting it right the first time.

14.8 Summary

The adoption of e-commerce will increasingly be achieved through:

- Commercial pressure to reduced data handling costs, time and queries.
- Business improvement initiatives that are reliant on effective transaction handling.
- Hub services increasingly supporting business processes.
- Simplified roll out and value-driven partnering.
- Confidence (and/or pressure?) resulting from progress made by competitors.

Realizing the full benefits of e-commerce in construction requires a strong relationship to be forged between business and technology. At the same time, the whole process needs to be driven by well-informed managers with clear sponsorship from senior management. Solutions should only be adopted that are flexible, scalable, technically competent and able to support ongoing innovation.

The operational experience of early adopters has confirmed the value proposition and shown that significant gains will become available as more and more companies embrace e-commerce.

For companies face in trying to accelerate the take up of e-commerce, perhaps the biggest challenge is in identifying and communicating a value proposition to all parties involved in the data exchange process. By ensuring benefits are shared across all parties, the industry will unlock the door to greater adoption and thereby dramatically reduce amount of time and money wasted in manually moving data between different computer applications.

References

EU Directive on Electronic VAT Invoices (2001/115/EC) http://eur-lex.europa.eu/smartapi/cgi/sga_doc?smartapi!celexapi!prod!CELEXnumdoc&numdoc=32001L0115&model=guichett&lg=en (accessed 13 September 2007).

Whittle, J. (2002) *eCentre^{UK} Presentation to CITE Annual Conference*, Birmingham, UK.

15 Assessment of e-Business Implementation in the US Construction Industry

Raymond R.A. Issa, Ian Flood and Bryce Treffinger

15.1 Introduction

The construction industry is one of the most people-intensive industries in the world. Construction is also known for its conservative attitude toward adopting new technology over the past decade. Traditionally, the construction industry is the last one to accept changes brought about by advancements in technology when compared to other industries. Today, as e-business has advanced beyond a buzz word, the construction industry has started to respond to it, by reforming the way it does business.

Based on US Department of Commerce data, the value of construction put in place in the United States in 2004 was $1 trillion, more than 8.5% of the gross domestic product of $11.7 trillion and it is expected to grow to $25 trillion by the year 2030 (Nelson, 2004). The total cost of replacement of real estate during the same period will exceed $21 trillion in the United States alone. These industry trends will force the construction industry to deliver a better product with enhanced customer involvement and satisfaction. Producing the best quality product and achieving the highest level of customer satisfaction requires all the team players to work jointly on their construction project. Internet-based tools will facilitate such collaboration among team players. This collaboration will improve the quality of the end product satisfying the client, and will also improve the efficiency of product development thereby satisfying the construction project team. e-Business initiatives have the potential to transform the construction industry and allow the construction industry to derive improved efficiencies out of e-business besides just cost cutting.

This chapter explores the extent to which the US construction industry has redefined its way of doing business based on e-business. It determines current e-business implementation solutions in the US construction industry and discusses the future implementation potential for various industry participants: general contractors, design firms, suppliers, and subcontractors in general. It focuses on the integration of e-business into construction project management systems by general contractors. The development of business transactions such as e-procurement, trade exchange, and business strategies such as customer relationship management (CRM), construction project collaboration, construction project management, enterprise resource planning (ERP), and knowledge/data management are discussed. The findings of the study presented in this chapter are also compared to the results of a survey completed by the authors in 2000 using the same survey instrument.

Thus the purpose of the study on which this chapter is based, was to conduct a survey to determine how much the construction industry has improved in its usage of e-business over a five-year period (2000–2005). The study uses data collected for a previous study conducted using companies selected from the 1999 Engineering News Record (ENR) Top 400 Construction Companies list and compares it to data collected in 2005 using the 2004 ENR Top 400 Construction Companies list. The businesses surveyed represented construction project management and construction services companies throughout the United States and varied in terms of annual revenues, workforce size, and geographical locations. At the same time the spread and use of communication technology in the construction industry is also documented since it is the basis upon which e-business applications are built.

15.2 US construction industry

15.2.1 Background

The effects of e-business on the construction industry in terms of its targets and changes and the improvements derived from its application in construction are shown in Figure 15.1 (Issa *et al.*, 2003). In general, e-business enables transactions to take place online, thus increasing the accuracy and efficiency of business transaction processing while optimizing business processes, condensing business cycle times, reducing cost, and improving customer service. It eliminates obstacles between corporate and business partners or customers. e-Business enables partners and customers to communicate and share information via the Internet, and it is used to serve customers, and provide the right information to the right people at the appropriate time. e-Business also creates market transparency for its customers, such as corporations, suppliers, partners. Figure 15.2 shows the relationships between various e-business initiatives in the architecture, engineering, and construction (AEC) industry.

Global construction projects come in different partnership formats such as joint ventures, outsourcing, and subcontracting or local representatives at the construction project location. This brings a new level of complexity to the industry: communication, different relationships, partnering, new markets, and global business standards and rules (Tucker, 1997). For example, Bechtel, which is one of the largest construction companies in the United States (The Engineering News Record Top 400 2004), exhibits

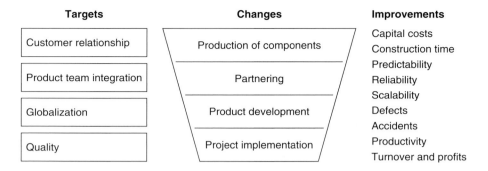

Figure 15.1 Effects of e-business on the construction industry

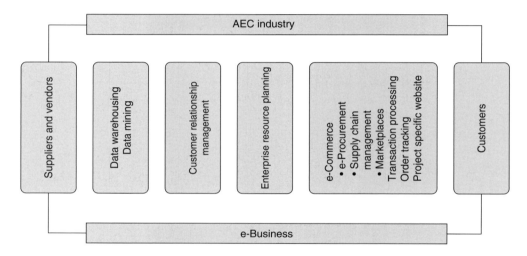

Figure 15.2 e-Business AEC industry relationships

a well-defined global structure. It has 41 000 employees working in 66 different countries. Bechtel booked $18.5 billion in new business for 2000 and had $18.1 billion in revenue (Bechtel, 2007).

15.2.2 Impact of e-business

e-Business can be defined as the conduct of buying and selling of goods and services as well as business communication and transactions over computer networks and through individual computers linked to the Word Wide Web (Zou and Seo, 2006). e-Business has become a worldwide phenomenon and is not different from ordinary business. It is where the use of Internet technologies, intellectual property, and customer superiority are combined and integrated with business activities which alter the traditional business model of operations (Chang and Ping Li, 2003). The potential use of *e*-business technologies applications in the construction industry include: e-marketing, e-selling/e-procurement of goods and services, e-collaboration, e-finance, and e-customer services and relations (Veeramani *et al.*, 2002).

Knowledge sharing

Knowledge sharing is recognized as a channel for the construction industry to address its need for innovation and improved business performance (Anumba *et al.*, 2005). However, frequently, organizations rely heavily on people and assume that they will transfer their learning and experiences to other employees, which can make organizations vulnerable if and when there is a high staff turnover. People-based knowledge transfer systems may not incur much cost commitment to organizations, but such approaches tend to be ineffective, unproductive, and expensive when compared with the loss of knowledge that is inevitable when employees leave the organization, resulting in possible difficulty in the case of expansion (Kamara *et al.*, 2002). Hence, companies whether large or small, whether public or private, from any industry, are using e-business as a means to organize business communications and improve success rates. e-Business has become increasingly popular over the past five years and with the advancement of information technologies it is becoming easier to adapt and use from any location. In addition, customers and users are more comfortable with the technology and use it regularly for personal business.

Owners have become more demanding and construction industry leaders want information available to them 24 hours a day, seven days a week. Electronic marketplaces allow the construction industry to find the most appropriate manufactured products as well as services that are being developed for, and being made available to, the construction industry. Reductions in the costs of operations, increased productivity, and improved customer satisfaction are being realized through the use of e-business strategies. The productivity improvements offered by newly developed e-business services can reduce and often eliminate unnecessary mistakes, time, energy, and costs, all the while keeping construction projects on track.

The construction industry has great potential in realizing the benefits of e-business. There is a great deal of opportunity especially in construction e-markets. Usage of the Internet can enable companies to more efficiently track construction projects, realize results faster, reduce risk, and hold parties accountable for their actions. In the past, the construction industry seemed to reject the Internet and all the new technologies it has to offer because of the industry's deep roots in traditional thinking. The Internet has the potential to rid the industry of inefficiency and cost. Because it has stuck to the old methods of regular mail, couriers, and faxes, the construction industry often gets a bad reputation for finishing construction projects late and over budget. However, construction industry leaders are apprehensive of implementing e-business because of the very traditional nature of the industry. It is an extremely people-oriented business. Relationships are established with suppliers and subcontractors that have lasted for years making the option of using a subcontractor found online that one has never met before, a turnoff for most traditional construction leaders. However, e-business still encompasses the people-oriented nature of construction because of its own need for people. Without the human interface, the systems would not succeed at all. e-Business requires an integrated alignment of technology, operation, strategy, structure, and human interaction in a continuously expanding network (Chang and Ping Li, 2003).

Benefits of e-business

The benefits realized from e-business are not limited to reduced costs, but also includes improved predictability, productivity, reliability, and scalability, ability to detect defects, improved levels of service, and extended market research. All these advantages as well as the development of software applications that allow users to get more for their money, are attracting more companies to engage in e-business (Issa *et al.*, 2003). For example, Swinerton & Walberg Builders cut change order turnaround time by more than half by using Bidcom.com, an online construction project management program (Fisher, 2000).

The construction industry is a multibillion dollar industry and any time and money saved on daily operations can quickly add up to large numbers. The construction industry claims that 60–80% of the total cost of operation, including capital, labor, materials, and transportation, is directly related to information management (Greissler, 2001). This information management pertains to everything from construction project scheduling to ordering materials to designing and coordinating construction and shop drawings. More time is actually spent on the business side of construction with sharing information than is spent onsite actually constructing the structure.

Business-to-business e-business in construction

The opportunity for the construction industry to take advantage of business-to-business (B2B) e-business and produce positive results for their company bottom line is significant. B2B connects customer, supplier, and partner applications, as well as business processes across the Internet. Supply chain partners can use B2B for shared planning as well as synchronized manufacturing, and distribution management. The main purpose of B2B is to automate business operations and information (Paper *et al.*, 2004).

Industries associated with construction, such as design and facilities management, or various infrastructure such as commercial buildings, manufacturing plants, roads, highways, public and private construction projects alike, together constitute a worldwide market of more than

$4 trillion a year (Cleveland, 2001). However, construction companies are not using this B2B e-business to their full advantage. A large industry such as construction should take notice of the growing technological trends and invest time and efforts into becoming more technologically savvy in order to create greater successes for construction companies. With such a large amount of money at stake in construction projects worldwide, companies could be using B2B e-business more to their advantage to streamline operations and increase company benefits.

The advantages B2B e-business can offer in terms of speed and cost savings need to be realized by the construction industry (Anumba and Ruikar, 2002). The construction process is much more complicated than most other industries because it is highly fragmented in terms of both participants and processes; it involves a variety of construction projects and the fact that every construction project is unique from owner to designer to construction project manager. This fact can make most standardized practices for other industries a nightmare for the construction industry. Because there are so many people involved in any construction process it can be difficult to process business transactions.

It is through 'straight-through processing' that the streamlining of the processes involved in construction can be achieved. The B2B process involves entering the necessary information one time into a system and allowing the system to take it through the process with minimal human interaction (Cleveland, 2001). A Web-centric system, or an integrated network of computer devices and information appliances that manages, stores, and distributes information through WWW specifications, can support collaborative environments more readily through a combination of Internet technologies. Among these technologies are HTML-based Web pages with Java Applets, Java Scripts, CGI Scripts, Databases, FTP, peripheral devices, as well as other new data formats (Rojas and Songer, 1999). The ongoing development of bcXML (Lima et al., 2003) and aecXML (IAI, 2006) has also started to facilitate the communication of product information and services across national and international boundaries, thus enhancing the prospects of successful deployment of e-business applications in construction.

Although construction project-specific Websites (PSWS) have been extremely useful with regards to construction team collaboration and the exchange of documents and information, they really do not have anything to do with B2B e-business and the selling and/or buying of construction items and materials for a certain construction project. PSWS merely promote the sharing of information for construction projects and nothing more. 'Without components, we are forced to revert to the traditional – manual – methods for identifying and quantifying the materials we need to purchase and erect. We inject a human right into the middle of our e-business transactions' (Cleveland, 2001). Dealing with components is what makes processes easier thus streamlining activities and speeding up processes.

Additional approaches were suggested by other researchers aimed at integrating e-business into the construction industry. Zhang and Tiong (2003) proposed a conceptual model called the integrated electronic business model. The model is used as a tool to compare the interaction and integration of construction processes, in the electronic business environment versus those in the traditional environment. Li *et al.* (2003) investigated the roles of Internet-based GIS in e-business systems in order to provide better services in location-based queries, business area analysis, and transportation analysis. They implemented an e-business system for construction material which they called construction materials exchange. Ruikar *et al.* (2006) described the development and implementation of VERDICT, a prototype to assess the e-readiness of construction companies in terms of their management, people, processes, and technology.

15.3 e-Business assessment survey findings

In order to determine the degree of implementation of e-business in the US construction industry an e-business assessment survey was created based on an earlier survey by Issa *et al.* (2003). The survey was distributed to a random sample of 91 companies out of the 2004 ENR Top 400 US Contractors ranked by gross annual revenue. The survey questions were designed to find out how far the construction industry respondents have advanced in the implementation of e-business applications, what type of companies use e-business transactions, and to what extent they use these applications. The survey is focused on the company size, geographical distribution, revenue, e-business transaction use, and e-business investments and future plans for e-business implementation. It is based on a combination of mail and Web-based responses from 20 corporations out of the 91 selected from the 2004 ENR Top 400 Construction Companies. The businesses surveyed represented construction project management and construction services companies throughout the United States and varied in terms of annual revenues, workforce size, and geographical locations.

The number of companies responding was 20 which constituted a return rate of 22%. A search of the literature suggests that a sample size of 20 or more is adequate (Ostle and Malone, 1988). In addition to the demographic information about the respondents, selected results of the survey are presented within the following four study areas: adoption of e-business, communication tools usage, e-business initiatives, and the respondents' prioritized goals for e-business.

15.3.1 Survey demographics

The demographics explored by the survey include the job function of the respondents, the size of their workforce, and the geographical distribution

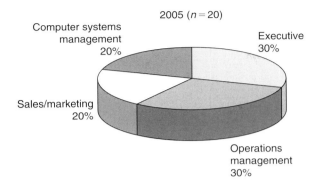

Figure 15.3 Respondents' job functions. (*Source*: 2005 figures, n = 20)

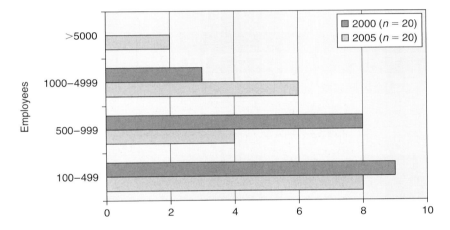

Figure 15.4 Respondents' workforce size distribution

of their operations. Figure 15.3 shows the distribution of respondents by job functions. The large number of operation managers (35%) and executives (25%) responding to the survey indicates that these groups are the most involved in their construction company's e-business decisions. The difference in workforce size distribution among the survey respondents is shown in Figure 15.4 with the majority of respondents having between 100 and 1000 employees.

15.3.2 e-Business implementation

The distribution of e-business applications implemented by the respondent companies is shown in Table 15.1 and in Figure 15.5 and is longitudinally compared with the results for the year 2000 results. Every company surveyed was involved with e-business applications, in one form or another. The most widely used e-business application in 2005 is that of

Table 15.1 Adopted e-business applications in construction and software source

e-Business initiatives	2000 results (n = 20)				2005 results (n = 20)			
	N	%	Source Internal	Source External	N	%	Source Internal	Source External
e-Procurement	8	40	3	0	6	30	2	4
Customer relationship management	6	30	3	0	8	40	7	1
Workflow	8	40	3	1	8	40	6	2
Supply chain management	5	25	2	1	3	15	3	0
Extranet/Intranet	13	65	7	1	12	60	11	2
e-Commerce	9	45	4	0	6	30	3	3
Knowledge Management/ Data warehousing	9	45	5	0	12	60	9	3
Internet infrastructure	12	60	5	1	10	50	8	2
Enterprise resource planning	5	25	4	0	7	35	6	1
Accounting/finance	11	55	5	0	11	55	7	4
Construction project collaboration	10	50	4	5	13	65	7	6
Construction project management	14	70	5	4	16	80	9	9
Digital exchange/auction	1	5	0	1	6	30	1	5
Wireless	0	0	0	0	8	40	7	1

Extranet/Intranet at 75% (15) and construction project management is close behind with a 70% (14) adoption rate. The survey results also show more adoption in the wireless category than in 2000. The survey in 2000 showed 0% adoption of any kind of wireless technology. Presently 45% (9) of the respondents are utilizing the wireless technologies in part because of technological advances as well as greater user familiarity with this technology. Accounting and finance tools are among the most implemented tools and are used a great deal at 65% (13) with most internally developed. Although more than 45% of companies required cell phones, which use wireless technologies, the respondents were most likely considering wireless connections for laptops on site and for wireless data exchange over the Internet via personal digital assistants (PDA). The combination of these three tools has increased the adoption of each tool by itself and explains their greater adoption rate in the construction industry. Among the least adopted applications even though they have great potential in saving money and streamlining operations are e-commerce, e-procurement, and supply chain management. The number of adoptions of these applications is sure to rise especially with the new

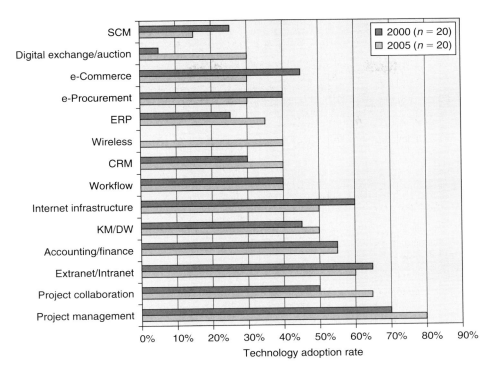

Figure 15.5 Adopted e-business applications in 2000 and 2005

generation of more computer savvy individuals accustomed to conducting business over the Internet who are getting ready to enter the construction work force. Some of the e-business initiatives shown in Table 15.2 showed a decrease from the year 2000 to the year 2005. Those decreases may be attributed to the fact that the respondents the surveys were different and the fact that most of these e-business initiatives have become more commonly accepted and were thus less noted by the respondents.

Figure 15.6 shows the distribution of the respondents' types of connection with their customers and suppliers. Although the survey indicated that the construction industry most often uses the old method of communicating by phone or in person there has been an increase in the use of email by respondents. The use of wireless technologies has increased from 5% (1) in 2000 to over 35% (6) in 2005. The use of public and private marketplaces is pretty low considering the recent surge in digital marketplace presence. The US construction industry is heavily dependent on personal relationships and seems to still work by word of mouth, asking around and using past suppliers and manufacturers for business operations. Wireless computing will allow the reorder of an item from a digital marketplace at a moment's notice should something go wrong with a current building material.

Table 15.2 Respondents' priorities

Business goals	2000 (n = 20)					2005 (n = 20)				
	Average	Minimum	Maximum	Rank	%=5	Average	Minimum	Maximum	Rank	%=5
Increased external communication	4.9	4	5	1	65	3.5	2	5	8	35
Increased internal communication	4.9	4	5	1	65	4.6	1	5	2	70
Enhanced customer relationships	4.9	4	5	2	60	4.6	2	5	2	70
Expansion of geographical opportunities	2.8	1	5	12	15	3.1	2	5	10	15
Innovation of product/services delivery	3.9	3	5	6	20	3.6	2	5	7	10
Shorter, accurate transactions	3.7	2	5	7	15	3.3	1	5	9	15
Transparent market	2.8	2	5	12	5	2.2	1	5	12	5
Expansion of partnership	3.5	2	5	10	15	3.0	2	5	11	5
Reduced capital costs	3.5	2	5	8	25	3.9	2	5	5	30
Reduced travel costs	3.5	2	5	8	10	3.5	1	5	8	20
Increased productivity and profitability	4.6	4	5	4	45	4.7	2	5	1	70
Increased predictability and performance	4.7	4	5	3	50	4.5	3	5	3	65
Reduced defects and accidents	4.1	3	5	5	30	4.2	3	5	4	55
Improved industry standards	3.2	1	5	11	5	3.8	2	5	6	25

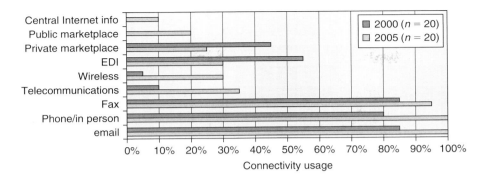

Figure 15.6 Suppliers, partners, and customers connectivity

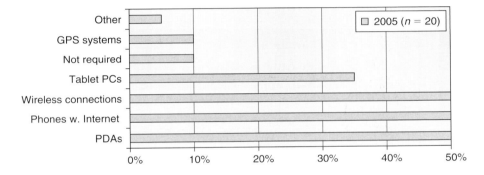

Figure 15.7 Computing devices supplied to respondents

Figure 15.7 shows the distribution of the type of wireless computing devices supplied to employees. With the apparent use of these mobile computing devices it is clear that the construction industry is embracing new technologies and looking for new ways to increase productivity. These devices simplify processes and streamline workflow which is exactly what the complex construction industry needs. After all in the construction industry time is money.

The distribution of e-business initiatives among the respondents' companies is shown in Figure 15.8. Current practices and future implementation in procurement, supply chain, transactions, e-commerce, project development, Intranet/Extranet, e-markets, order tracking, partnering, and communication were all covered to determine the industry's needs. Ninety percent (90%) of the respondents have implemented e-commerce initiatives and have plans for future e-commerce initiatives within their organization. Project Development and Intranet/Extranet tools were the most used and/or are slated for greater use. The greater use of Intranet/Extranet at 60% (12) versus 55% (11) in 2000 was no surprise considering the high response rate earlier with regards to e-commerce applications.

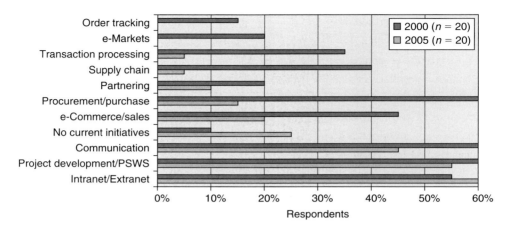

Figure 15.8 e-Business initiatives in the construction industry

Communication was the next most frequently picked response among the initiatives with a 10% increase from 50% (10) in 2000 to 60% (12) in 2005. With a constant need to keep in touch with suppliers, clients, and everyone involved in the project, communication is extremely important to construction industry leaders. These tools are important to maintain and increase productivity while reducing costs, staying on schedule and keeping up with the current rapid market growth.

15.3.3 Prioritized goals of the construction industry

A comparison of the distribution of the respondent companies' goals between 2000 and 2005 is shown in Figure 15.9 and in Table 15.2. Customer relationships are a most important concern for leading construction companies ranking second in 2000 and moving to the number one goal in 2005. Increased productivity and increased predictability is a close second, moving from fourth and third respectively in 2000 to trying for second in 2005. It is not clear as to whether or not the respondents or the construction industry are familiar with B2B exchange because 35% (7) of the respondents did not know whether their company had participated in B2B exchange, 40% (8) indicated no participation, and another 20% (5) indicated participation in this application. These results show that B2B exchange has not been implemented as much as expected, which may be due in part to the process itself, which involves entering information into a system and allowing the system to take it through the process with minimal human interaction (Cleveland, 2001). These problems are mainly due to the complexity and fragmentation of the construction industry.

Figure 15.10 shows the obstacles that the respondents perceived that the construction industry is faced with when implementing e-business. Since e-business is so new to the construction industry there are few cases

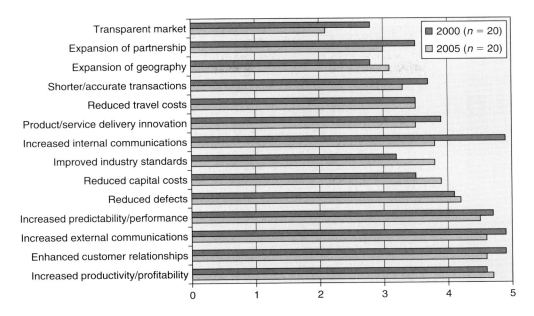

Figure 15.9 Respondents' prioritized goals

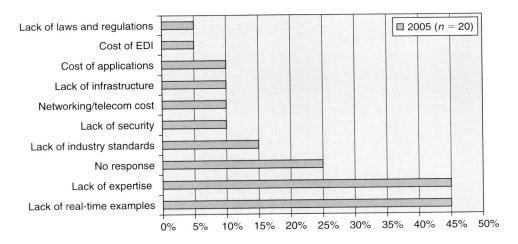

Figure 15.10 Reasons for not implementing e-business in construction

of companies who have implemented e-business applications with real-time documented results. In addition, the construction industry is wary of implementing too much e-business at one time because of its lack of technical expertise. e-Business is still relatively new to construction, which indicates that there is going to be a lack of expertise in this area across the board. However, with the increase in computer literacy among

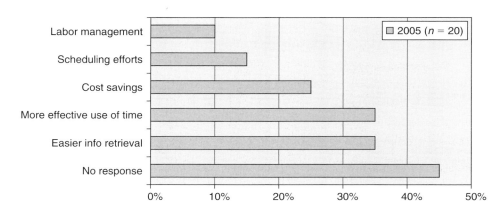

Figure 15.11 Benefits realized in construction from e-business

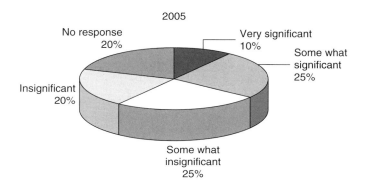

Figure 15.12 Impact of e-business on construction revenue. (*Source*: 2005 figures)

the workforce and the increased expectations of customers, the implementation of e-business is expected to be on the rise.

The benefits that the respondents have experienced with the implementation of e-business applications are shown in Figure 15.11. The ability to retrieve project information with ease and to allow for more effective use of time are the greatest benefits noted by the respondents, with 40% (8) feeling these were the benefits their company was realizing. These benefits really go hand-in-hand, because with the better use of time comes cost savings which is another benefit the respondents perceived. Figure 15.12 shows the 2005 respondents' perception of the impact of e-business on revenue. Most of the respondents felt that their e-business adoption was somewhat significant with 35% (7). A higher proportion of respondents 45% (9) felt there was some sort of impact on revenue from e-business implementation. The four that did not respond were those same individuals that felt they were not fully knowledgeable in their company's e-business efforts.

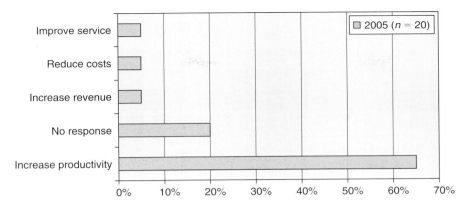

Figure 15.13 Perceived benefits of e-business implementation in construction

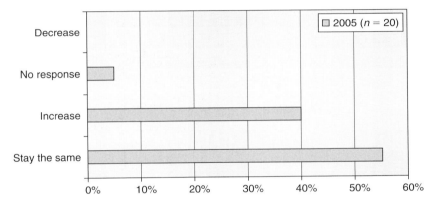

Figure 15.14 Expected future spending on e-business implementation in construction

Figure 15.13 shows that an overwhelming number of the respondents have implemented e-business in the hope of increasing productivity, thus leading to more effective use of time, and in turn cost savings. The construction industry is based on getting things completed as fast as possible with the least amount of steps. It is evident in this response that this 'getting it done' attitude is what is driving the decisions on e-business adoption.

As illustrated in Figure 15.14, the respondents show no signs of wanting to rid their operations of e-business implementation. Either way they plan on continued spending on e-business in the construction industry. The companies realize that e-business implementation is benefiting business operations and that continued support for e-business is crucial to construction business success.

15.4 Conclusions

Companies that have begun to re-evaluate their technology needs and usage, that have evaluated new Web-based solutions, and that have developed, refined, selected, and prioritized a set of solutions will be in a good position to realize considerable cost savings, to increase operating efficiencies and to improve customer satisfaction and profitability. e-Business is about the commitment and capability of companies in various industries to utilize digital technology and to enhance customer satisfaction across their business functions, thus changing their way of doing business from a traditional company-centric stand-alone paradigm to a new network-leveraged synchronized paradigm (Chang and Ping Li, 2003). The results of the survey administered in 2005 indicate that the most widely used e-business application was that of Extranet/Intranet followed by project management. Project Development and Intranet/Extranet tools were the most used and/or are slated for greater use in 2005. Lack of real-time examples and expertise were the most often cited obstacles that the construction industry is faced with when conducting e-Business. The most often cited benefits were the user's ability to retrieve project information with ease and the more effective use of their time. Future surveys should focus on finding out whether the current trends toward enhanced integration, collaboration, and interoperability will result in even greater improvements in productivity, time and schedule management, and cost control and reduction.

References

Anumba, C.J., Egbu, C. and Carrillo, P. (2005) *Knowledge Management in Construction*. Blackwell Publishing, Oxford, UK.

Anumba, C.J. and Ruikar K. (2002) Electronic commerce in construction – trends and prospects. *Automation in Construction*, **11**(3) April, 265–275.

Bechtel Corporation (2007) <http://www.bechtel.com> (accessed 12 March 2007).

Chang, S.S. and Ping Li, C.H.P. (2003) How to succeed in e-commerce by taking the higher road: Formulating e-commerce strategy through network building. *Competitiveness Review*, **13**(2) June, 34–46.

Cleveland, A.B. (2001) B2B in the construction industry: Putting first things first. *Leadership & Management in Engineering*, **1**(1) January, 56–57.

Fisher, S.E. (2000) Can construction adapt to online markets? *InfoWorld*, **22**(20) May, 38.

Greissler, R. (2001) Building data bridges. *Roads & Bridges*, **39**(2) February, 42–45.

International Association for Interoperability (2006). <http://www.iai-na.org/aecxml/> (accessed 23 May 2007).

Issa, R.R.A., Flood, I. and Caglasin, G. (2003) A survey of e-commerce implementation in the US construction industry. *Journal of Information Technology in Construction*, **8** May, 15–28.

Kamara, J.M., Augenbroe, G., Anumba, C.J. and Carrillo, P.M. (2002) Knowledge management in the architecture, engineering and construction industry. *Journal of Construction Innovation*, **2**(1) January, 53–67.

Li, H., Kong, C.W., Pang, Y.C., Shi, W.Z. and Yu, L. (2003) Internet-based geographical information systems for e-commerce application in construction material procurement. *Journal of Construction Engineering and Management*, **129**(6) November/December, 689–697.

Lima, C., Stephens, J. and Böhms, M. (2003) The bcXML: Supporting eCommerce and knowledgemanagement in the construction industry. *Journal of Information Technology in Construction*, Special Issue on eWork and eBusiness, **8** October, 293–308.

Nelson, A.C. (2004) *Toward A New Metropolis: The Opportunity to Rebuild America*. The Brookings Institution, Washington, DC.

Ostle, B. and Malone, L. (1988) *Statistics in Research*, 4th edition, Iowa State University Press, Ames, Iowa.

Paper, D., Pedersen, E. and Mulbery, K. (2004) An e-commerce process model: Perspectives from e-commerce entrepreneurs. *Journal of Electronic Commerce in Organizations*, **1**(3) July, 28–47.

Rojas, E.M. and Songer, A.D. (1999) Web-centric systems: A new paradigm for collaborative engineering. *Journal of Management in Engineering*, **15**(1) January/February, 39–45.

Ruikar, K., Anumba, C.J. and Carrillo, P.M. (2006) VERDICT – An e-readiness assessment application for construction companies. *Automation in Construction*, **15**(1) January, 98–110.

The Engineering News Record 2004 US Top 400 Construction Companies (2004). <http://www.enr.com> (accessed 23 May 2006).

Tucker, R. (1997) Emerging global opportunities in construction. In: *Proceedings ASCE Construction Congress IV*, 1–3 October, Philadelphia, USA, 1–8.

Veeramani, R., Russel, J.S., Chan, C., Cusick, N., Mahle, M.M. and Roo, B.V. (2002) State-of-practice of ecommerce application in the construction industry. *CII Research Report*, 180–211.

Zhang, N. and Tiong, R. (2003) Integrated electronic commerce model for the construction industry. *Journal of Construction Engineering and Management*, **129**(5) September/October, 578–585.

Zou, P.X.W. and Seo, Y. (2006) Effective applications of e-commerce technologies in construction supply chain: Current practice and future improvement. *Journal of Information Technology in Construction*, **11**, 127–147.

16 Concluding Notes

Chimay J. Anumba and Kirti Ruikar

16.1 Introduction

This chapter concludes this book and highlights a number of issues relating to e-business implementation in the construction industry. It starts with a brief summary of the various sections of the book, emphasizes the benefits of e-business to the construction industry, and discusses the enablers and challenges in its practical implementation. The last section of the chapter explores some of the future directions in e-business in construction and identifies promising research areas.

16.2 Summary

The focus of this book has been on e-business adoption in the construction industry. This is in recognition of the numerous benefits that the industry stands to reap from the adoption of e-business tools in its processes. It also recognizes and addresses the need for human and organizational changes in the industry to facilitate e-business implementation.

The introductory part of the book (Chapters 1–5) served as a general introduction to the subject of e-Business in Construction. Chapter 1 introduced the subject and the contents of the book while Chapter 2 focused on the fundamental principles of e-business and its evolution. Chapter 3 provided the construction context for e-business, exploring the trends and issues in the industry that make the adoption of e-business an imperative. Chapter 4 explored in some detail the need for construction sector organizations to assess their readiness for e-business implementation, and presented a model that has been specifically developed for this purpose. Chapter 5 extends this further by discussing the necessary infrastructure for e-business implementation across multi-disciplinary teams, such as exist in construction projects.

The second part of the book (Chapters 6–10) focused on presenting a range of technological solutions designed to support the implementation of e-Business in Construction. Chapter 6 explored the evolving role of extranets in construction e-business, while Chapter 7 presented an agent-based approach to construction e-business and illustrated the key concepts through a description of the features of a system for the specification and procurement of construction products. Chapter 8 addressed the use of electronic hubs (e-Hubs) to facilitate e-business between major clients and groups of small- and medium-sized enterprises (SMEs). In Chapter 9, the important role that Web Services and XML play in e-business is discussed and a prototype system, E-Union, presented. Chapter 10 presented a futuristic perspective by discussing the potential for Semantic Web-based e-business, which has the capacity to enable enhanced interoperability between e-business systems.

The third part of the book (Chapters 11–13) covered the socio-technical aspects of e-business in construction. Chapter 11 addressed the issue of trust and trust building, which are considered vital ingredients in construction e-business. The focus of Chapter 12 was the legal issues inherent in e-business. It provided an insight into the key risks that construction supply chain members need to be aware of and make provision for. The importance of knowledge management in construction e-business was discussed in Chapter 13.

The fourth part of the book (Chapters 14 and 15) focus on the industrial perspective on construction e-business implementation. Chapter 14 draws on case studies to present the UK perspective while Chapter 15 presents the US perspective based on an industry survey.

16.3   Benefits of e-business in construction

The chapters in this book have, to varying extents, highlighted the benefits of e-business in construction. In particular, Chapter 2 discussed these benefits from the perspectives of the key members of a construction supply chain. While many of the benefits are generic to the application of e-business in any industry sector, some are particularly relevant for construction sector organizations. It is useful to reiterate these benefits here:

- Reduced advertising and marketing costs.
- Provision of company information (products and services) through a Web presence.
- Easy access to target audiences from the construction sector, and transparency with customers.
- Quicker access to construction-related information.
- Up-to-date product and industry information.
- Simplified procurement processes.
- Cost savings through disintermediation and quicker product comparison in terms of price, functionality, and quality.
- Better management of the construction project delivery process.
- Easier access to project information from anywhere at anytime.
- Faster transaction times.
- Improved transparency in the exchange of project information.
- Time savings in the communication of project information.

- Savings on project cost.
- Streamlined construction business processes.
- Reduced paperwork.
- Reduced re-keying of information (thereby reducing errors).
- Wider market reach for construction product manufacturers.
- Lower transaction costs.
- Reduced staffing requirements.
- Shorter procurement cycles.
- Decreased inventory levels for product manufacturers.
- Provision of information on demand which promotes its better use/reuse.
- Connection to operations across organizational boundaries.
- Enlargement of the span of effective control and co-ordination.
- Improvement in the quality of decision-making processes.
- Enhanced communication and collaboration between supply chain members.

The effectiveness of the e-business implementation is critical to the realization of the above benefits within individual firms and/or the whole supply chain. In this regard, it is important to review some of the key considerations in construction e-business implementation.

16.4 Considerations in construction e-business implementation

16.4.1 Key considerations

Construction organizations and supply chains that intend to adopt e-business need to address a number of key issues to ensure that they maximize the benefits outlined above. It is imperative that organizations and supply chains undertake a readiness assessment, as discussed in Chapter 4, to ensure that they have all the critical ingredients in place to ensure a successful e-business implementation. Some of the main considerations in e-business implementation include the following:

- The availability of a supportive infrastructure for e-business implementation within an individual firm and/or across the supply chain. This includes both the IT and non-IT systems necessary for effective e-business.
- The existence of an organizational framework and policies that support both individuals and teams, and enables the e-business implementation to be well managed.
- The need for a clear business strategy that outlines an organizationís objectives with regard to electronic business interactions with clients and other supply chain members.

- The readiness of an organization (from the management, processes, people, and technology perspectives) for effective adoption of e-business.
- Management buy-in is critical to e-business implementation, as it is required to define the business goals and ensure the alignment of all other factors in achieving these.
- The appropriateness of the ICT infrastructure for enabling the communication and exchange of business information in a collaborative environment.
- It is important the people in an organization or supply chain are an integral part of the e-business implementation, as their abilities, attitudes, feelings, relationships, and training can have a profound impact on the success or failure of an e-business implementation.
- The scope for business process improvements both to increase efficiency as well as to accommodate new e-business tools and business process.
- The provision of training to enable staff and supply chain members to fulfil their roles within the e-business environment.
- The institution of appropriate procedures and policies for quality assurance in e-business processes.
- The availability of appropriate technologies to facilitate information exchange and knowledge sharing.
- Use of common hardware and software platforms to ensure the seamless exchange of business information. The use of standard and proven information and communication technologies may be helpful in this regard.

16.4.2 Barriers to e-business implementation

There are many barriers to the implementation of e-business in construction, and consideration needs to be given to overcoming these in order to ensure success. Some of the main barriers that need to be overcome include:

- The fragmentation and traditional adversarial relationships between supply chain members.
- The lack of trust between supply chain members.
- The lack of access to an appropriate e-business infrastructure (i.e. the Internet, telecommunication systems, power supply, etc.) for many SMEs and/or supply chain members in developing countries.
- Concerns over the security of electronic transmission of commercially sensitive information over public networks.
- Adherence to traditional, tried and tested but outmoded business methods.
- The lack of an appropriate regulatory framework for e-business transactions.
- The conservative nature of the construction industry and associated inertia in the uptake of new technologies and business processes.
- The transient nature of construction project teams, which inhibits investment in long-term e-business systems.

- Difficulties in quantifying the return on investment, particularly for SMEs.
- Low levels of awareness and understanding of the available technologies, opportunities, and benefits of e-business.

These barriers can be addressed in a variety of ways but by far the most promising approaches include the following:

- Improvements in education and training for both new entrants and established practitioners in the construction industry
- Increased use of collaborative project delivery systems that encourage the building of trust between supply chain members.
- Provision of incentives by clients and government agencies to encourage the conduct of business transactions by electronic means.
- Improved communication of the benefits of e-business by those organizations and supply chains that have reaped demonstrable benefits.
- Improvements in the existing infrastructure for e-business, including an appropriate regulatory framework, better encryption and security protocols, etc.
- The proactive adoption of established and emerging information and communications technologies (e.g. project extranets, Web Services, XML, Web-based portals, and collaboration systems, etc.) that facilitate construction e-business.
- The establishment of electronic hubs that facilitate the involvement of SMEs (which constitute more than 80% of construction sector organizations) in construction e-business.

16.5 Future directions

e-Business changes at a rapid rate as new technologies constantly emerge and new functions are added to existing technologies. Such technological developments impact on various facets of an organization, including its culture, processes, and resources. This requires that companies adopt a vigilant approach and establish proactive measures to respond to these changes. It is impossible to predict, with any credibility, the direction that e-business in construction will take in the future. This section will, therefore, simply draw on a number of industry trends and emerging information and communication technologies to outline some of the issues that will have an impact in shaping the future:

- The issue of trust is central to e-business in construction. This is now being increasingly recognized and a number of research projects are being undertaken. These are expected to deliver models, frameworks, processes, and tools that will facilitate trust building.

- The advent of the Semantic Web, which is expected to facilitate more knowledge sharing and collaborative working, will have a major impact on construction e-business. This requires the development of appropriate ontologies for construction e-business.
- Significant changes are required in construction business processes if the full benefits of e-business are to be realized. However, the exact changes will differ from one company to another. It is therefore important that systems for effective change management are put in place within organizations and across supply chains. There is scope for further studies on the most appropriate ways to institute changes at individual, team and organizational levels, and (crucially) at the interfaces between these.
- e-Business implementation will reshape the inner workings of construction supply chains. Work is needed to understand how best to integrate e-business processes into supply chain management.
- There is a need for the development of appropriate metrics for the evaluation of the benefits of e-business over conventional approaches. However, it must be recognized that not all benefits can be quantified.
- The potential for mobile commerce based on the use of mobile devices, such as mobile phones and PDAs (Personal Digital Assistants), has long been recognized. The increasing functionality of these devices and the supporting wireless communication networks make this a potential growth area.
- The trend towards embedding intelligence in many household and office appliances, some of which have the capacity to be connected to the Internet, will result in these devices being participants in e-business transactions. The possibilities are potentially huge but difficult to predict.
- Procurement of goods and services is now possible through intelligent autonomous software agents (see Chapters 7 and 9). The user informs the software agent about the purchasing tasks to be performed, and the software agent then acts in a proactive manner by accomplishing the tasks. The growth in Web Services is expected to increase the automation of many purchasing tasks (for both goods and services).

There is clear evidence from the contents of this book that e-business has much to offer construction sector organizations. The complexity associated with the delivery of construction projects by a transient project team made up of individuals/teams from a variety of organizations makes the implementation of e-business challenging. However, this also makes the successful implementation of e-business in construction highly rewarding for all members of the construction supply chain.

Index

Accela.com, 9
Administration-to-Administration (A2A), 9
Administration-to-Business (A2B), 9
   see also Business-to-Administration (B2A)
Administration-to-Consumer (A2C), 10
   see also Consumer-to-Administration (C2A)
AEC-specific e-business, see agent-based systems
aecXML, 156, 253
Agency Law, 218
agent-based systems, 104, 106–7
   APRON, 109–14
      architecture, 116–17
      conceptual design, 114–16
   current context, 104–6
   software agents, views on, 106–7
   vision for, 107–9
agent communication, 115, 116
Amazon.com, 9, 26, 72
Antitrust Acts, 202
APEC's E-Commerce Readiness Initiative, 45
application services provider (ASP), 158
applications, of e-business
   and end-user construction companies, 32
      end-user processes, 36–9
      product reviews, 32–6
APRON, 109–14, 118
   implemented APRON prototype, 116–17
   conceptual design, 114–16
Architecture, Engineering and Construction (AEC) sector, 65, 81, 92, 101, 104, 105, 249, 250
ARROW (Advanced Reusable Reliable Object Warehouse), 110, 111

barriers, to e-business, 16
   construction-specific, 18–20
   generic, 17–18
Basic Collaboration Platform (BCP), 137
bcXML, 117, 156, 158, 180, 253
BEACON model, 46, 47
Bechtel, 249, 250

benefits
   of construction collaboration technologies, 86
      intangible benefits, 90
      quasi-tangible benefits, 87–9
      tangible benefits, 87
   of e-business, 252
      to construction organizations, 17
Bidcom.com, 252
BIW Technologies, 100, 109, 110, 111
blogs, 8
Building Information Model (BIM), 156
BuildNet, 71
business process improvement, 26
business process reengineering (BPR), 26, 51
   for e-business, 26–7
   representative BPR model, 27–31
business rules, 153, 244–5
Business-to-Administration (B2A), 9
   see also Administration-to-Business (A2B)
Business-to-Business (B2B), 7–8, 123, 124, 128, 150, 195, 205
   e-business, in construction, 252–4
   e-trading marketplaces, 150–51
Business-to-Consumer (B2C), 9, 124, 128, 150, 195
   see also Consumer-to-Business (C2B)

c-negotiators, 217
Center for Research in Electronic Commerce (CREC), 109
CITE (Construction Industry Trading Electronically), 236, 237
client tier, 53, 54, 152, 153, 162, 163–4
collaborative project planning, 132, 134–5
collaborative working, 92
   individual resistance to, 92–3
   industry resistance to, 95–6
   inter-organizational resistance to, 94–5
   intra-organizational resistance to, 93–4
COMMA, 176
common data standard, 236–7
   to hub-centric environment, 237–9
'community of practice', 183, 184, 227

construction collaboration technologies, 81–3, 137
  benefits, 86
    intangible, 90
    quasi-tangible, 87–9
    tangible, 87
  human aspects, 91
    collaboration technologies, resistance to, 96
    collaborative working, resistance to, 92–6
    human/technology issues, managing, 97–9
  issues, 85
  KM system for, 188–91
  moving beyond collaboration, 99–101
  technologies for, 137–8
  uptake, 83–6
    barriers, 84–5
construction industry, e-business in, 14, 23
  adoption, 24–5
  applications and end-user construction companies, 32
    end-user processes, 36–9
    product reviews, 32–6
  barriers, to e-business, 16
    construction-specific, 18–20
    generic, 17–18
  benefits, 15–16, 17, 251
  BPR for, 26–7
    representative BPR model, 27–31
  effects, 250
  goals, 260–63
  initiatives in, 260
  legal issues in, 211–18
  next generation web technologies, 168–9
construction materials exchange, 254
construction projects, 14, 18, 25, 32, 33, 76, 82, 83, 86, 167, 168, 169, 170, 248, 249, 253
construction-specific barriers, to e-business, 18–20
Consumer-to-Administration (C2A), 10
  see also Administration-to-Consumer (A2C)
Consumer-to-Business (C2B), 8, 9
  see also Business-to-Consumer (B2C)
Consumer-to-Consumer (C2C), 8–9, 67
contract, definition, 72
Coprocure.com, 128
'creative abrasion', 227
customer relationship management (CRM), 10, 168, 248
Cyber Business Centre, 124–5

data, definition, 74
data fidelity, 175
Data Protection Act 1998, 19
database management system (DBMS), 54, 55, 152
database tier, 53, 54, 55
Detrimental reliance, *see* promissory estoppel
digital rights management (DRM), 204

disintermediation, 9
dispositional trust, 198
document sharing systems, 170–71
Document Type Definitions (DTDs), 157
door supplier selection process, 30

'early-design-process ontology', 190
eBay, 9, 67, 69, 127, 195, 203, 204
e-Bid.co.uk, 127
e-business, definition, 6, 250
e-business marketplace infrastructure, 67, 68
e-business models, 11–14, 26, 73
e-business readiness assessment protocol, 45, 230, 231
  primary phase, 230, 231–2
  secondary phase, 230, 232
  tertiary phase, 230, 232
e-business systems
  for construction products procurement, 151
    architecture, 152–3
    E-Union concept, 154–5, 160–64
    limitations, 153–4
    need for, 149–51
    standardization, 155–7
    types, 152
    Web Services model, 157–60
e-buying and selling, 127–8
EC (European Commission), 179
e-COGNOS, 180
e-collaboration, 128–9
e-commerce
  definitions, 6–7
  in construction, 235
    background, 235–6
    case studies, 242–6
    common data standard, 236–7
    first adopters, 241–2
    hub-centric environment, common standard to, 237–9
    internet, role of, 237
    issues and benefits, 239–41
  e-Hubs' role in, 123
    concept, 124–5
    core services, 145
    engineering e-Hub, 132–40
    engineering services, 140–44
    problems and challenges, 144–5
    value-added services, 146–7
    Web Services, 125–31
  trust building, 196
    concept, of trust, 196–7
    issues, 197–9
    lifecycle, issues in, 200–206
    practices in, 206–8
    risks, 199–200

e-Construct project, 179
e-construction infrastructure framework, 66–7
    challenges, 69
        information liquidity, 74
        knowledge management, 74
        legal, 71–3
        managerial and organizational, 73–4
        multi-disciplinary, 75–7
        technological, 70–71
    importance, 67–9
EDIFACT (Electronic Data Interchange for Administration Commerce and Transport), 236
Egan Report, 81, 104, 105
e-Hubs' role, in e-commerce, 123
    concept, 124–5
    core services, 145
    engineering e-Hub, 132
        functional architecture, 133–40
        role, in project preparation process, 132–3
    engineering services, 140–44
    problems and challenges, 144–5
    value-added services, 146–7
    Web Services, 125
        collaborative Web Services, 125–7
        examples, 129–31
        services offered, by e-Hubs, 127–9
        taxonomy, 126
Electronic Commerce Bill, 19
Electronic Data Interchange (EDI), 158, 208
electronic document management (EDM), 37, 38, 53
Electronic Union (E-Union) concept, 151, 154–5
    Web Services prototypical implementation, 160
        multi-tier products catalogue architecture, 162–4
        products catalogue searching model, 161–2
eLEGAL, 137, 144, 204, 218
element, in XML document, 156
e-marketplace, 116, 124
engineering e-Hub, 132
    engineering services, 140–44
    functional architecture, 133–40
    problems and challenges, 144–5
    role, in project preparation process, 132–3
Engineering News Record (ENR), 69, 249, 254
Engineering Service Providers (ESPs), 132, 141, 201
Enterprise Applications Integration (EAI), 137
Enterprise Resource Planning (ERP), 217, 249
e-readiness, 39, 44–5
    end-user e-readiness, 47–53
    see also organizational readiness
e-Steel.com, 123
e-tendering, 23, 100, 101
e-trading marketplaces, 150–51
e-transaction fulfilment, 128

E-Union Web Services, prototypical implementation, 160–64
    multi-tier catalogue architecture, 162
        client tier, 163–4
        E-Union member tier, 162
        E-Union tier, 163
    products catalogue searching model, 161–2
    system architecture, 160
explicit knowledge, 74, 229, 230
eXtensible Markup Language (XML), 126, 151, 155, 156–7, 174, 175, 177, 237, 244
extranets, in e-business
    construction collaboration technologies, 81
        benefits, 86–90
        defining, 81–3
        human aspects, 91–9
        moving beyond collaboration, 99–101
        uptake, 83–6

faces, of e-business
    business and financial models perspective, 10
    commerce, 11
    relationships, 10
    responsiveness, 11
FIATECH, 70
FreeMarkets.com, 127
functional architecture, of e-Hub, 133, 138–40
    collaboration, technologies for, 137
    collaborative project planning
        theoretical basis, 134–5
    process management
        workflow management system for, 135–7
    supporting technology, 135
functional process improvement, *see* business process improvement
fundamentals, of e-business
    barriers, to e-business, 16
        construction-specific, 18–20
        generic, 17–18
    e-business enablers, 15–16
    e-commerce and e-business definition, 6–7
    faces, 10–11
    models, 11–14
    taxonomy, 7
        Administration-to-Administration, 9
        Business-to-Administration, 9
        Business-to-Business, 7–8
        Business-to-Consumer, 9
        Consumer-to-Administration, 10
        Consumer-to-Consumer, 8–9
    trends, 14–15

Ganttproject, 137, 144
Gartner Advisory Group, 6, 10

GCC, 128
General Contractor (GC), 212, 216, 217
Graphical User Interface (GUI), 137

'Hub Alliance', 238
'Hubs', 238, 239, 243, 244
human/technological issues management, 97
   buy-in, 97–8
   cost, 99
   exchange standards, agreement, 98
   selection, 97
   timing, 97
   training, 99
Hypertext Markup Language (HTML), 54, 127, 153, 161, 163, 172
Hypertext Transfer Protocol (HTTP), 151, 157, 161

IDS, 125
If It Works, It's Not AI: A Commercial Look at Artificial Intelligence Startups, 119
ifcxml, 117, 156
impact, of e-business, 250
   benefits, 252
   Business-to-Business, in construction, 252–4
   knowledge sharing, 251–2
Industry Foundation Classes (IFC), 156
information, definition, 74
information and communication technologies (ICTs), 23, 44, 167, 168, 169, 179, 188, 222, 270
information channel (IC), 27, 28, 29
information liquidity, 74, 76
information overload, 108, 171–2
Initiator agents, 217
institutional trust, 198
intangible benefits, of construction collaboration technologies, 90
integrated e-business infrastructure framework, 65, 69
   e-construction infrastructure framework, 66–7
      importance, 67–9
   information liquidity challenges, 74
   knowledge management challenges, 74
   legal challenges, 71–3
   managerial and organizational challenges, 73–4
   multi-disciplinary challenges, 75–7
   skeleton for
      information infrastructure, 67
      legal and regulatory infrastructure, 66
      managerial and organizational infrastructure, 66
      technological infrastructure, 66
   technological challenges
      connectivity and reliability, 71
      inadequate software, 71
      interoperability, 70
      security, 70–71
integrated electronic business model, 254
Integration DEFinition language 0 (IDEF0), 27
intelligent components (I-components), 28–9, 111
International Alliance for Interoperability (IAI), 70, 156
Internet, 6, 14, 15, 17, 26, 171, 173, 174, 213
   organizational challenges, 226–8
   role, 237
Internet-based electronic products catalogues, 151
interoperable construction products catalogues
   Web Services model, 157–60
IQ Net Readiness Scorecard, 46–7
IT Construction Forum, 84
ITNET, 123

Kelly's law of networks, 67, 68
KM system
   for construction collaboration and e-business, 188–91
knowledge, definition, 74
knowledge-based business (k-business), 222, 230–33
knowledge gap identification process, 224
knowledge management (KM), for e-business, 180, 222
   challenges, 74
   in context, 223–4
   knowledge asset employment
      by construction organizations, 229–30
   organizational challenges, 226–8
   perspective, 224–6
   readiness assessment protocol, 230–33
knowledge management system, 178, 183, 189
knowledge modelling, 175–6
'knowledge workers', 92

Latham report, 81
leadership, 227
legal challenges, of e-business, 71–3
legal issues, in construction e-business, 211
   legal risks, types of, 212
      agency, 216
      agent negotiation, 217
      attribution, 215–16
      authentication, 215–16
      contract formation, validity and errors, 212–13
      jurisdiction, 213–14
      non-repudiation, 215–16
      prelude, 216
      privacy, 214–15
      variation, 217–18
London Olympics 2012, 9
Loughborough University, 109

Macromedia Dreamweaver and Fireworks, 54
managerial and organizational challenges, of
　　e-business, 73–4
many-to-many liquidity, 150
McGraw Hill Sweets Product Library, 111
Mediator Agents, 217
meta-data, 178, 179
meta-models, 179
Metcalfe's law, 67
Microsoft ASP.Net framework, 160
Microsoft Internet Information Services, 160
Microsoft SOAP, 160
Microsoft SQL Server 2000, 160
Microsoft SQLXML, 160
middle tier, 53–4, 55, 152, 153, 162, 163
Mortenson Co., Inc.
　　vs. Timberline Software Corp., 75
Mosaic's readiness assessment tool, 45
MS Internet Explorer, 55
multi-disciplinary challenges, of e-business, 75–7
MySQL, 55

nano-publishing, 8
Netscape, 55
Network for Construction Collaboration Technology
　　Providers (NCCTP), 82, 84, 87, 88, 89, 90
'Networked Readiness Index', 45
next generation web technologies, 167
　　construction context, 168–9
　　semantic web, 169
Nitze, P, 50
Node A253 product supplier selection process, 29–31

Onestopcaradvice.co.uk, 12
ontologies, 174–7
OntoShare, 183
ONTOWEB, 176
OntoWise, 183–8
operational e-commerce, 236
Organization for Economic Cooperation and
　　Development (OECD), 7
organizational knowledge assets, 229
organizational readiness, for e-business, 42
　　end-user, case study, 56–60
　　to launch k-business, 230–33
　　methodology, 44–5
　　readiness assessment models, review of, 45–7
　　VERDICT, 47
　　　background, 47
　　　implementation, 53
　　　management, 49–50
　　　operation, 55–6
　　　people, 51–2

　　　process, 50–51
　　　system architecture, 53–5
　　　technology, 52–3

paradigm shift, 51
partnership environment, in Web Services model, 159
Peer-to-peer, 8
personal digital assistants (PDAs), 256, 271
PHP, 54, 55
PI e-Hub, 128
PICS (Platform for Internet Content Selection), 174
Podcasting, 8
poor content aggregation, 172
privilege management infrastructures (PMIs), 208
process automation, 51
'process workers', 92
Procurement Module, 117
procurement process, 240
product reviews, 32–6
project document management and distribution
　　using Semantic Web, 181
　　　deployment scenario, 182–3
　　　OntoWise, 183–8
　　　overview, 181–2
project extranets, 42, 82, 84, 179, 189, 190
Project Information Exchange (PIX) Protocol, 98
Project Management Body of Knowledge, 134
project planning model (PPM), 134, 135, 138, 139, 140
project-specific ontology, 190
project-specific Websites (PSWS), 253
promise, definition, 212
promissory estoppel, 75
Purchase Order, 35

quantifiable benefits, *see* tangible benefits, of
　　construction collaboration technologies
quasi-tangible benefits, of construction collaboration
　　technologies, 87–9

RACE, 46
rapid application development (RAD), 44–5
rationalization, of procedures, 51
readiness assessment models, 45–7
'Readiness Gap', 233
Request for Proposal (RFP), 212, 215
Resource Description Framework (RDF), 174, 177,
　　190, 191
'Restatement of Contracts', 212
Ross, D.T., 27

SADT (structured analysis and design technique), 27
SCALES (Supply Chain Assessment and Lean
　　Evaluation System), 46

SciQuest.com, 123
screen scrapping, 172
SEKT project, 180
Semantic Web, 168, 169
   e-business, in construction sector
      KM system, for construction collaboration and e-business, 188–91
      project document management and distribution, 181–8
   evolution, in construction sector, 178–81
   key concepts, 172–4
   knowledge representation and manipulation, 177–8
   need for, 169–72
   ontologies, 174–7
small- and medium-sized enterprises (SMEs), 18, 129, 135
SOAP (Simple Object Access Protocol), 125, 157, 160, 163
SofTech, Inc., 27
software agents, 72, 106, 119, 178, 216, 271
   vs. standard software, 217
   views on, 106–7
Standard Generalized Markup Language (SGML), 156
static data, 243
stovepipe system, 172
structured information, 152
'style sheets', 244
Sun Java 2 Platform, 158
supplier products database, 31
survey demographics, 254–5
SurveyMonkey, 9
Surveys.com, 9

tacit knowledge, 74, 229
tangible benefits, of construction collaboration technologies, 87
taxonomy, of e-business, 7
   Administration-to-Administration, 9
   Business-to-Administration, 9
   Business-to-Business, 7–8
   Business-to-Consumer, 9
   Consumer-to-Administration, 10
   Consumer-to-Consumer, 8–9
'team-profile ontology', 190
technological challenges, of e-business, 70–71
Technology Adoption Life Cycle, 37, 38
three-tier architecture, of web-based application, 54, 152–3
Timberline Software Corp.
   vs. Mortenson Co., Inc., 75
time-bound negotiation agents, 218
Timestamp.com, 208
traditional tendering, 100

trust and trust building, in e-commerce, 195
   concept, of trust, 196–7
   issues, 197–9, 200
      contract execution and post-contract fulfillment, 205–6
      contract negotiation, 204–5
      partner finding, data searching and contract negotiation, 203–4
      user registration, 201–2
   practices, 206
      business approaches, 206–7
      technological approaches, 207–8
   risks, 199–200
trust building model, 207
trusting beliefs, 199

Uniform Electronic Transaction Act (UETA), 72, 216
Uniform Electronic Transmission Act, 71
Uninterrupted Power Supplies (UPS), 71
Universal Description, Discovery and Integration (UDDI), 125, 157, 158
unstructured information, 152
URLs (Uniform Resource Locators), 177
US construction industry, 4, 248
   background, 249–50
   e-business assessment survey, 254
      demographics, 254–5
      e-business implementation, 255–60
      goals, of construction industry, 260–63
   impact, of e-business, 250
      benefits, 252
      Business-to-Business, 252–4
      knowledge sharing, 251–2
US Department of Commerce, 248

ValiCert, 207
VERDICT (Verify End-user e-Readiness using a Diagnostic Tool), 254
   background, 47
   end-user case study, 56
      average scores summarization, 57–8
      evaluation, 58–60
      radar diagram, 58
      summary report, 58
   implementation, 53
   management, 49–50
   operation, 55–6
   people, 51–2
   process, 50–51
   system architecture, 53–5
   technology, 52–3
Verisign, 207
VHS (Video Home System), 236

video casting, 8
ViewSum text summarization tool, 185

Web Services, of e-Hub, 125
   collaborative Web Services, 125–7
   examples, 129–31
   services offered by, 127
      e-buying and selling, 127–8
      e-collaboration, 128–9
      e-transaction fulfilment, 128
   taxonomy, 126
Web Services and aecXML-based e-business system
   for construction products procurement, 149
      architecture, 152–3
      E-Union concept, 154–5
      E-Union Web Services, prototypical
         implementation, 160–64
      limitations, 153–4
      need for, 149–51
      standardization, 155–7
      types, 152
      Web Services model, 157–60
Web Services model
   of interoperable construction products catalogues, 157–60
Workflow Management System (WfMS), 135–7
World Wide Web Consortium (W3C), 172, 173, 174
WSDL (Web Services Description Language), 125, 157, 158
WSIL (Web Service Inspection Language), 125

XML document
   parser, 157
   schemas, 116, 157, 175